ROUTLEDGE LIBRARY EDITIONS: COLD WAR SECURITY STUDIES

Volume 8

BURDEN-SHARING IN NATO

BURDEN-SHARING IN NATO

SIMON LUNN

Routledge
Taylor & Francis Group

LONDON AND NEW YORK

First published in 1983 by Routledge & Kegan Paul Ltd

This edition first published in 2021
by Routledge
2 Park Square, Milton Park, Abingdon, Oxon OX14 4RN

and by Routledge
605 Third Avenue, New York, NY 10017

Routledge is an imprint of the Taylor & Francis Group, an informa business

© 1983 Royal Institute of International Affairs

All rights reserved. No part of this book may be reprinted or reproduced or utilised in any form or by any electronic, mechanical, or other means, now known or hereafter invented, including photocopying and recording, or in any information storage or retrieval system, without permission in writing from the publishers.

Trademark notice: Product or corporate names may be trademarks or registered trademarks, and are used only for identification and explanation without intent to infringe.

British Library Cataloguing in Publication Data
A catalogue record for this book is available from the British Library

ISBN: 978-0-367-56630-2 (Set)
ISBN: 978-1-00-312438-2 (Set) (ebk)
ISBN: 978-0-367-63011-9 (Volume 8) (hbk)
ISBN: 978-1-00-311186-3 (Volume 8) (ebk)

Publisher's Note
The publisher has gone to great lengths to ensure the quality of this reprint but points out that some imperfections in the original copies may be apparent.

Disclaimer
The publisher has made every effort to trace copyright holders and would welcome correspondence from those they have been unable to trace.

Chatham House Papers · 18
Burden-sharing in NATO

Simon Lunn

The Royal Institute of International Affairs

Routledge & Kegan Paul
London, Boston and Henley

The Royal Institute of International Affairs is an unofficial body which promotes the scientific study of international questions and does not express opinions of its own. The opinions expressed in this paper are the responsibility of the author.

None of this paper was written while the author was employed by the North Atlantic Treaty Organization and the views expressed are purely his own.

First published 1983
by Routledge & Kegan Paul Ltd
39 Store Street, London WC1E 7DD,
9 Park Street, Boston, Mass. 02108, USA and
Broadway House, Newtown Road,
Henley-on-Thames, Oxon RG9 1EN

Printed in Great Britain by
Billing & Son Ltd, Worcester

© *Royal Institute of International Affairs 1983*

No part of this book may be reproduced in
any form without permission from the
publisher, except for the quotation of brief
passages in criticism.

Library of Congress Cataloging in Publication Data
Lunn, Simon.
Burden-sharing in NATO.
(Chatham House papers: 18)
Includes bibliographical references.
1. North Atlantic Treaty Organization.   I. Royal
Institute of International Affairs.   II. Title.
III. Title: Burden-sharing in N.A.T.O.   IV. Series.
UA646.3.L83 1983   355'.031'091821   83-4497
ISBN 0-7100-9233-4

Contents

	Acknowledgments	vi
1	Introduction	1
2	The historical background	7
3	The Soviet invasion of Afghanistan	19
4	The alliance dilemma	29
5	The Reagan Administration	33
6	Burden-sharing	53
7	Problems and prospects	62
	Tables	80
	Notes	83

Acknowledgments

This paper forms part of a wider Chatham House research project on US foreign policy and European interests which has been funded by the Leverhulme Trust.

I should like to acknowledge the assistance and encouragement of Professor Lawrence Freedman. My thanks also go to Miss Louise Rosewarne and Miss Louise Orrock, who have been responsible for typing a number of drafts. — S. L.

1 Introduction

Over the past three decades, the question of burden-sharing has made regular appearances on NATO's agenda. At issue is how the resource burden of NATO's security arrangement should be calculated and distributed fairly among the allies. The issue has normally been put on the agenda by the United States in the belief that it is carrying a disproportionate share of the burden, and this has been the case in recent years. On close examination the problem of burden-sharing soon moves beyond specific questions of how to measure and compare the defence expenditure of the various countries, and on to much more fundamental questions. How is the burden to be defined? Does it include just the allocation of forces to the NATO commands, or does it also comprise efforts to promote Western interests outside of the NATO area? And, in a wide definition, should foreign aid be included as well as more military-related activites? To what extent should West Europeans feel obliged to pay for the exercise by the United States of what it takes to be its global role, and to what extent does the size of the US contribution reinforce its position of leadership within NATO? Is the burden-sharing issue essentially a surrogate for a much more difficult argument over American doubts about the loyalty and determination of its allies and allied doubts about the widsom of recent American policy?

As will be seen in this paper, burden-sharing has been about much more than who pays for what in NATO. In consequence the debate has been frustrating, because the wider issues have often not been recognized, and inconclusive, because the particular question of burden-sharing, with all its specific complexities, is an unsatisfactory way of dealing

Introduction

with these wider issues. As will be argued in this study, it will only be by addressing directly the basic security relationship between the United States and Western Europe that the issue of burden-sharing can be laid to rest.

NATO was founded in April 1949. It was set up to provide a system of collective security for its members and to guarantee a condition of stability in the North Atlantic area. Initially the nature and scope of 'the burden' involved in ensuring North Atlantic security was clearly identifiable: it combined a counter to the perceived menace of Soviet military power with the rebuilding of an economically devastated Europe. But there was a question of means and priorities. Europe could not recover economically and at the same time devote the resources to defence that were considered necessary to counter Soviet military strength. The immediate solution to this dilemma was found through the Atlantic partnership and the harnessing of American military and economic potential to Europe. The United States provided economic assistance through the Marshall Plan, and military protection through its participation in NATO.

In the early years of the transatlantic partnership, the United States shouldered a large proportion of the burden of Western security. For Americans, the weight of this burden was somewhat lightened by the knowledge that an economically sound and military secure Europe was in their interest: it would provide markets essential for America's expanding economy, and it would constitute a vital bulwark for the containment of communism. From the outset, however, the commitment wavered uneasily between the realization that assisting in the defence of Europe was in America's own interest and a deep-rooted apprehension that America's effort in this respect should not allow its allies to contribute less than they should for their own defence.

As Europe has become more prosperous, this instinctive suspicion of European 'free riding' in the defence of alliance interests has intensified. Many Americans criticize the fact that European economic growth has not been accompanied by a commensurate increase in the resources that the European nations allocate to defence, a criticism that has led to the accusation that the United States cares more about defending Europe than the Europeans themselves. This perception provides the basis for the burden-sharing debate and explains why the issue is mainly

Introduction

an American preoccupation.

While American dissatisfaction on this issue has surfaced sporadically throughout the history of the alliance, it has normally been contained within the reconciliatory nature of the alliance framework. However, during the last two years Congressional frustration with its allies has intensified, and the question of burden-sharing has re-emerged with considerable force. It now represents a major divisive factor in alliance politics. Yet, as a problem, the issue of burden-sharing is insoluble. No matter how comprehensive the analysis, comparisons that depend on selected economic indices or static force levels can do no more than describe a part of the total picture.

The most appropriate comment on the usefulness of the burden-sharing exercise is the fact that despite the conclusion of thoroughly researched Pentagon reports to the effect that the allies are contributing their fair share of the collective burden, Congressional criticism continues, as if the reports did not exist. From this, it is evident that such criticism represents an instinctive and subjective reaction to the unbalanced nature of the Atlantic relationship, rather than a rational and balanced assessment of the totality of that relationship.

Despite the intractability of the burden-sharing issue, it remains a fact of alliance life. The politics of the Atlantic relationship necessitate that efforts are made to assess and compare the defence efforts of member countries. This exercise can serve a constructive function in that it provides a general benchmark against which individual members can reassess their respective contributions in relation to the total effort. Given the pressure for defence to assume a lower priority against competing social demands, there is a certain value in a degree of gentle prodding in order to ensure continuity of effort.

However, an excessive preoccupation with burden-sharing can have negative consequences. It can encourage nations to concentrate on areas that provide visible proof of burden-sharing but little in the way of additional collective capability, rather than on more mundane measures which may lack visibility but contribute to general effectiveness. Most important, however, an attempt to apply common standards of performance runs counter to the diversity which is inherent in the structure of the alliance. NATO is a voluntary association of free and independent states. Countries can be encouraged and cajoled into

Introduction

making greater efforts for the collective good, but, at the end of the day, defence expenditure is a national responsibility based on the support of national electorates. This support can be generated only from within; it cannot be mandated from outside.

The debate over burden-sharing is a product of the structure of the Atlantic relationship. A recurring irritant, it has never by itself proved decisively damaging to the cohesion of the alliance. But it has provided a convenient foundation of criticism on which other disputes can flourish. The residual resentment over burden-sharing exacerbates, and is exacerbated by, disagreements in other areas. In this respect, the importance of burden-sharing as a transatlantic issue is directly related to the prevailing climate in the Atlantic relationship.

There has rarely been a period when NATO has been free from some kind of transatlantic friction. The geostrategic differences between the United States and Europe have provided fertile ground for dissent and disagreement. In this sense, the history of the alliance is a study in the continuity of diversity. The same issues reappear with monotonous regularity: the credibility of the American nuclear guarantee for extended deterrence, the need for better conventional forces, the levels of American troops in Europe, American urgings to do more and European pleadings of budgetary constraints, the perennial emphasis on doing better with existing resources, differences over economic relations with the East, and European regional preoccupations as opposed to America's global interests.

Whenever serious disputes have occurred, these have been reconciled by an alliance decision-making process in which a willingness to compromise has been the predominant characteristic. Agreement in principle has generally been produced, even if implementation has later been found wanting. The flexibility of the alliance structure and its ability to 'fudge' difficult and controversial issues has allowed NATO to accommodate very different national views and preoccupations. This process has also been helped by the fact that for many years NATO has operated in the shadows. A lack of public attention has allowed governments greater latitude in dealing with politically sensitive questions and in accepting compromises which could create domestic problems.

Within this decision-making process, the United States has always played a dominant role. It has been in the forefront of most alliance

Introduction

initiatives, both nuclear[1] and conventional. Since most of these initiatives have involved doing additional tasks or spending more money, their reception by the Europeans has normally been one of caution and, not infrequently, opposition. European reticence towards new initiatives has a variety of causes: an unwillingness to move away from the *status quo*, or to find more resources, or to change long-standing force plans. For Europeans, continuity in force-planning has always been an extremely important factor. Defence-planning cycles in most European countries are well-established, and changes of government in Europe rarely produce major changes in defence policy. Moreover, European parliaments normally play little more than a oversight role and rarely affect defence policy in a meaningful way. In the United States, by contrast, not only does Congress wield a significant influence on defence decisions, but incoming administrations frequently feel obliged to assert their own particular brand of leadership on the alliance and consequently seek corresponding adjustments in alliance policies. The desire to provide distinctive leadership can result in suggested shifts in policy which, to say the least, are disruptive.

American leadership in NATO stems not only from America's military contribution, but also from the sheer weight of analytical resources that the United States can muster for the solution of any problem. Mere quantity of expertise does not guarantee quality of analysis. But it does afford the United States an advantage in the alliance decision-making process. This aspect, and the dynamics of the NATO decision-making process, explain why allied reactions to American initiatives are frequently ambiguous. The consensus principle of NATO meetings exerts a strong pressure on countries not to break ranks. Within the collegiate atmosphere of ministerial meetings, it is often easier for smaller allies to acknowledge new proposals or endorse them 'for further study' rather than object or criticize. Occasionally allied opposition or criticism of a specific proposal is muted because the Atlantic agenda of disagreement is already overloaded, and defence may be the least painful decision to give way on. However, decisions taken within the spirit of collectivity are inevitably exposed to the realities and constraints of domestic politics. Consequently there is a tendency for Europeans to endorse American initiatives, but to implement these initiatives only where they are consistent with existing national plans.

5

Introduction

The flexibility of the alliance decision-making process has enabled NATO to survive a number of crises. This resilience has led many veteran alliance-watchers to conclude that the common needs and mutual interests that underpin the Atlantic relationship will always prevail over spasmodic tensions and differences. Yet there are signs that this presumption may be losing its validity. During the last two years, transatlantic tension between the United States and Europe has increased to the extent that many observers now question whether the United States and Europe continue to share the same interests and objectives. Indeed, many predict that the current differences represent the first stage of a change that is historically inevitable — the disengagement of the United States from Europe and a major transformation in the structure of Western security.

This study examines some of the factors that have contributed to this interpretation of the state of the Atlantic relationship. It begins by demonstrating that the issue of 'burden-sharing' is inherent in the structure of the alliance and has been a source of conflict from the start. The role of the crisis over how to respond to the Soviet invasion of Afghanistan brought the issue to the fore with renewed intensity. It took on a new dimension with the growing preoccupation with a Soviet challenge outside of Europe, and this placed further strains on NATO's decision-making process. The resultant disagreements, which have become particularly intense during the Reagan Administration, are then described before moving on to an analysis of how the alliance has attempted to cope with this divergence in perspective. The study concludes by considering a number of alternative approaches to bridging the gap that currently exists between the United States and Europe concerning Western security arrangements.

2 The historical background

The early years

From its origins, the North Atlantic Alliance was faced with the almost insuperable task of accommodating within a single framework of collective security basic asymmetries in geography and power. This meant, in turn, reconciling different attitudes to global and regional security. Reconciliation of these differences was effected through a framework which, because of its loose structure, appeared to offer a high degree of insurance at relatively low cost. Right from the start, however, there was no agreement either on how much collective insurance was needed, or on how its payment should be apportioned. As this brief history will show, many of the problems that occurred in the early years still test the alliance. The lines remain the same even if in certain instances the characters have changed roles.

The Atlantic partnership was initially conceived for the dual purpose of ensuring the economic restoration of a war-ravaged Europe and protecting Europe from the perceived threat of Soviet military power. The American leadership of the time saw grave military and political implications in the formation of communist governments in Central and Eastern Europe following the war, and was concerned lest economic and social chaos in Western Europe should lead to communist infiltration. Its immediate concern was for economic recovery rather than military security. Economic assistance was channelled through the European Recovery Programme, and security assistance through the formation of NATO in conjunction with a military assistance programme. American motives for these policies were mixed. The reconstruction of the

The historical background

European economies was necessary before they could take up the role of trading partners for the United States. Also, as the only country to have emerged from World War II better off, there was a sense of moral obligation to assist the recovery of those allies who had borne the brunt of the fighting and suffering. Enlightened self-interest and altruism apart, America's leaders recognized that it was in America's interests that Europe should become economically viable and remain outside the Soviet sphere of influence. Yet there was also a strong determination that any assistance, whether economic or military, should be relatively temporary. Americans saw their assistance as an eventual contribution to a strong, preferably integrated, Europe that would be able to look after itself. American credit would put Europe back on its feet in order that American involvement could be reduced to a minimum.

In the initial years, therefore, the major burden was economic recovery. Although the military potential of the Soviet Union constituted a serious threat, few considered that an invasion was imminent. Thus, American policy emphasized the primacy of economic recovery: 'Economic recovery is a prior necessity: therefore, the size of European forces must be such that they do not interfere with recovery. And it looks as though they will continue to be quite small for some time.'[1] If there was an implicit acceptance that the Europeans' contribution to their own security would be constrained by the economic situation, there remained nevertheless in American utterances more than a hint of the traditional American suspicion of Old World commitments and a desire to ensure that the Europeans would eventually look after themselves. Even when the outbreak of the Korean war had convinced American planners that a Soviet conventional attack on Europe was a possibility, Congressional voices cautioned the extent of any additional military commitment to Europe.

The announcement by President Truman in late 1950 that he intended to make 'substantial increases'[2] to American forces stationed in Western Europe was a significant development, unprecedented for the United States in peacetime, and it provoked a major controversy in the United States. Former President Herbert Hoover voiced his strong opposition to the American policy of aid to Europe. Arguing that the continental allies were mainly responsible for their own defence, he asserted that the United States, rather than concentrating on Europe, should strengthen its air and naval power, rearm Japan and stiffen its Pacific Front. The issue was vigorously debated in Congress during the early months of 1951, and joint hearings on the subject were held by the Senate Foreign Relations Committee and the Senate Armed Services Committee, an unusual combination which signalled the gravity of the question before them. Simultaneously, an extended discussion began on the floor of the Senate which has become known in US history as the 'Great Debate'. The end product of this Senatorial exercise was Senate Resolution 99, approved on 4 April 1951.[3]

Resolution 99 approved the initiative taken by the executive branch in regard to the dispatch of substantial additional armed forces to Western Europe and for the appointment of a US Supreme Allied Commander. It also manifested the belief of the Senate that the West European allies of the United States should make their fair and substantial contributions to the allied defence effort, that the US Joint Chiefs of Staff should certify that the European allies were making a realistic effort on behalf of European defence, that European ground forces should make the necessary contribution to allied ground forces, and provision should be made to utilize the military resources of Italy, the Federal Republic of Germany and Spain.[4] Congress was already signalling its concern that there should be no 'free ride' in Western defence.

European concerns during this period were similar to those of the United States but with different degrees of emphasis. There was common agreement that economic recovery should come first. However, European governments tended to place greater emphasis on the scale of the Soviet threat. Concerned over the implication for their economies of providing adequate security against Soviet military power, they were anxious to secure a permanent American commitment, particularly the protection of American nuclear weapons. American military strength offered the protective mantle under which they could repair their economies, and the formation of NATO provided the guarantee of this involvement. From the beginning, perceptions of the need for the North Atlantic Alliance were somewhat different. Americans saw it as a means to create a more viable and independent Europe which would mean a diminished American involvement; Europeans viewed it as a means of ensuring American involvement.

The issue of Europe doing more was omnipresent. But doing more

The historical background

for what? Since motivations for creating the alliance were different, it was difficult to establish firm priorities that would have any chance of support. The perception of a Soviet threat was very real, especially after the Korean war, but the importance of economic recovery continued to be the main preoccupation. The problem was not only one of perception; it was also one of structure. NATO was not organized in a way that would have permitted an effective allocation of tasks and responsibilities commensurate with the resources of individual members. Lacking supranational authority, it was able to impose no more than a superficial degree of coordination among the varying defence efforts of its members. This structural problem has continued to thwart attempts to impose a more rational and unified approach to the NATO force posture. Then, as now, the final criterion in deciding national contributions was what could be afforded rather than what was required. Occasionally, the invisible hand of collective responsibility has functioned via the Annual Defence Review to modify a nation's defence plans. Members aiming to reduce their contributions have been persuaded to think again. However, the gap that separates the capabilities required from the resources available has never been satisfactorily bridged. If it was difficult to persuade nations to do more, it was equally difficult to get them to make a better use of their collective resources. Despite the evident duplication inherent in the national force structures, no one was willing to sacrifice the concept of balanced forces for a more effective common defence.

Before the outbreak of the Korean war, an approximate division of labour was proposed that minimized the US continental commitment. According to Secretary of State Dean Acheson, the defence efforts of the parties to the alliance were to be based on a 'logical and practical division of labour' by which 'Each member of the Alliance will specialize in the kinds of forces and the production of weapons for which it is best suited and which will fit into a pattern of integrated defence.'[5] The general features of such a division of labour were outlined by General Omar Bradley. Following the principle that 'the man in the best position and with the capability should do the job for which he is best suited', the division would be that,

First, the United States will be charged with the strategic

The historical background

bombing. We have repeatedly recognized in this country that the first priority of the joint defence is our ability to deliver the atomic bomb.

Second, the US Navy and the Western Union naval powers will conduct essential naval operations, including keeping sea lanes clear. The Western Union and other nations will maintain their own harbour and coastal defence.

Third, we recognize that the hard core of the ground power in being will come from Europe, aided by other nations as they can mobilize.

Fourth, England, France and the closer countries will have the bulk of the short-range bombardment and air defence. We, of course, will maintain the tactical air force for our own ground and naval forces, and United States defence.[6]

The outbreak of the Korean war on 25 June 1950 profoundly altered these assumptions and provided the impetus for the first serious attempt to create ground forces sufficient to withstand a massive Soviet conventional attack upon Western Europe. The United States launched a full rearmament programme, which quadrupled American defence expenditures within three years, and President Truman announced substantial increases in American forces stationed in Europe. The European allies announced plans for increasing their defence forces, periods of military service and military expenditures. The United States also announced that henceforth aid to Europe would be directed primarily towards rearmament efforts, rather than to economic expansion.[7] Most significant, the North Atlantic Council agreed that a 'forward strategy' would be adopted in Europe, and that 'any aggression should be resisted as far to the East as possible'.[8] It was recognized that, as well as requiring far greater resources, this strategy would call for an integrated force under a centralized command, and consequently General Dwight Eisenhower was appointed Supreme Commander of NATO forces in Europe in December 1950.

During 1951 and 1952, NATO forces in Europe, then put at fifteen divisions, grew in number and in capability. Four new American and two additional British divisions arrived in West Germany, supplemented on a smaller scale by additional French units and a Canadian brigade.

The historical background

However, this build-up was not deemed sufficient, and in February 1952 the North Atlantic Council, meeting in Lisbon, approved the so-called Lisbon force goals of 50 combat-ready divisions and 4,000 aircraft.[9] The adoption and subsequent non-fulfilment of these goals symbolized an inherent problem for the alliance: that the military perception of the threat far outstripped the political willingness of nations to make the necessary resources available, and that nations tended to promise more than they could afford. The inspiration afforded by collective decision-making was soon dissipated when ministers returned to the realities of domestic politics and resource constraints.

During the early 1950s the United States, suspicious of European intentions, continually urged the Europeans to do more. The Europeans for their part pleaded social and economic constraints and sought to perpetuate the American commitment as a means of avoiding the full impact of looking after their own security. Unable to devise an appropriate burden-sharing formula or to produce the resources considered necessary, NATO fell back on attempts to make better use of the resources that were available. In the December 1952 communiqué, NATO ministers 'directed that more emphasis should be given to increasing the effectiveness of the Alliance and the units necessary for their support, rather than to the provision of greater numbers, to the extent that resources were not available for both tasks'.

As the largest, most wealthy and militarily most capable member, it was inevitable that the United States would dominate the alliance, and most initiatives have tended to flow from Washington. Throughout the history of the alliance, the United States has led the way in military measures designed to increase the amount of resources available or to improve the effectiveness of their use. Initially, by enabling Europe's recovery and by setting an example, it aimed its efforts at increasing the European contribution, since it was hoped that this would lead to reciprocal decreases in the American commitment. Unfortunately, however, the fact that the United States was always required to set the example produced the opposite effect to that desired and tended to reinforce the American investment in Europe.

By 1954, NATO had accepted that it would not be able to match Soviet conventional forces. The United States was unwilling to undertake such an expensive task by itself, and the Europeans preferred

The historical background

economic reconstruction to increased expenditure on military forces. In the words of two American authors, 'NATO was thus born with a psychological "complex" about conventional forces.'[10] Its inability to build a credible conventional defence force was demonstrated by its failure to fulfil the Lisbon goals and was consistently emphasized by various SACEURs (Supreme Allied Commander, Europe). Because of this conventional inadequacy, the alliance came to rely on the use of nuclear weapons.

American superiority in nuclear weapons appeared to make the threat of nuclear retaliation a cheap and efficient solution to the problems posed by Soviet conventional superiority. Confidence in this approach did not last long; it began to dissipate as the Soviet Union developed and deployed its own range of nuclear weapons, including battlefield systems. As a consequence, the alliance has implemented over the years a number of doctrinal and technical measures designed to improve the credibility of its strategy of nuclear deterrence. As will be seen, none of these measures has solved the underlying problems of extended nuclear deterrence.

By the late 1950s, American exasperation with the European defence contributions had largely abated. This was due, first, to a change in strategy that stressed greater reliance on nuclear weapons and, second, to the accession of the Federal Republic of Germany to NATO, which to an extent alleviated the pressures on the other members. However, the issue soon returned, when the Kennedy Administration sought, in the early 1960s, to raise NATO's conventional capabilities in order to reduce the previous dependence on nuclear weapons. Furthermore, a balance-of-payments problem across the Atlantic (which was starting to develop in the 1950s) became serious, and the United States began to press for 'offset' payments to cover the foreign exchange costs of US troops based in Germany. This issue came to a head in 1966 and was settled by agreements concluded in April 1967.

Initially an economic issue, the kind and amount of German offset became linked to a central issue of policy and security – the number of US troops in Germany. The issues raised by this question have never been totally resolved and are central to current discussions of burden-sharing: Is the presence of American troops in Germany in America's interests or are they there to do Europe a favour? What do they actually

The historical background

cost and how much would be saved if they were withdrawn? And, finally, what contributions should the host country make? It should be noted that while Britain followed America's example in obtaining offsets, none of the other nations stationing troops on Federal territory has enjoyed similar arrangements.

The European contribution to Western defence continued to attract considerable attention in the United States during the late 1960s and early 1970s. The Vietnam experience, the French withdrawal from the integrated military structure of NATO in 1966, and US economic problems combined not only to cause the Johnson Administration to put pressure on the Europeans to increase their defence efforts but to diminish support in the US Congress for the US overseas troop commitment in general. This produced a strong Congressional movement, led by Senator Mike Mansfield, to cut US forces in Europe.

The Mansfield amendments

The Senator from Montana introduced the first of the Mansfield Resolutions on 31 August 1966. In this Resolution the Senate was asked to resolve that 'A substantial reduction of US forces permanently stationed in Europe can be made without adversely affecting our resolve or ablity to meet our commitment under the North Atlantic Treaty.'[11] The Senator persisted with this Resolution from 1966 until 1973. However, despite substantial support in Congress, his proposals were always defeated.[12]

In assessing this period in terms of the overall US—European relationship, it should be noted that Senator Mansfield's arguments were based not on frustration or vindictiveness towards the allies, but rather on the simple view that Europe was now rich enough to look after itself. He argued that in view of European economic well-being and the general relaxation of tension between East and West, a US military presence was redundant and unnecessary—especially since this presence was aggravating America's economic problems. He never succeeded in his aim, however, because the Administration was consistently able to make a sufficiently persuasive case that, for one reason or another, withdrawals were not appropriate, that bringing American troops home would not result in savings, and that a strong and basically unaltered

The historical background

American presence in Europe continued to be in America's interest. The most significant development in checking possible Senate action was the start of negotiations between NATO and the Warsaw Pact on the mutual reduction of forces in Central Europe (MBFR). Hence the immediate pressure for troop withdrawals disappeared. However, the basic sentiments to which the Mansfield proposals appealed remained beneath the surface, easily accessible and exploitable whenever tensions appeared within the alliance.

Concern over the balance-of-payments deficit had not disappeared entirely, and the Nixon Administration probed various ways of getting the allies to pay for American troops, mainly in terms of offset deals, trade or monetary concessions. These, however, met with little success. The allies objected to the prospect of American troops becoming 'little more than mercenaries' in Europe, arguing that the American troop presence was in America's interest as well as Europe's. A combination of events in the mid-1970s caused Congress to ease the pressure for American troop reductions in Europe. The ongoing Vienna MBFR negotiations constituted a permanent constraint. The US Administration argued that the chances of getting the Warsaw Pact countries to reduce their forces would be undermined if the United States reduced unilaterally. In addition, reports of Warsaw Pact force improvements tended to weaken the case for Western troop reductions. On the financial front, the US balance of payments improved considerably in 1975, lessening pressure from that quarter.

Congressional interest in the alliance did not disappear, however. The 1970s saw several Congressional initiatives aimed at achieving a better harmonization of force improvements and, by making American actions conditional on a European response, encouraging the European allies to do more. As an illustration that Congress's concern with the balance-of-payments issue had not completely dissipated, the Jackson/ Nunn Amendment to the Department of Defense Authorization Act 1974 required that the European allies offset the balance-of-payments deficit incurred by the United States as a result of the 1974 costs of stationing US forces in NATO Europe. Failure to offset would have resulted in automatic reductions in US force levels. On 2 June 1975, President Ford, in the final quarterly report to Congress on the European offset effort, reported that the deficit had been more than

The historical background

fully offset and that the troop reduction provision of the Jackson/Nunn Amendment would not have to be implemented. This was effectively the last fling of the offsets issue, although the question of American stationing costs has persisted in a different form.

During this period, Congress also required that US forces be streamlined and, in an amendment to the DoD Authorization Act for 1975, that the US support forces in Europe be cut by 18,000, authorizing the Secretary of Defense to increase combat personnel in Europe by an equal number. In a further amendment to that Act, Congress placed increased emphasis on interoperability and standardization, and called for the preparation of a study of the costs of NATO's past failure to standardize. The so-called Culver/Nunn Amendment of the 1976 Department of Defense Authorization Act required the Secretary of Defense to report to Congress the initiation of any procurement action of any major new system which was not consistent with NATO standardization efforts. The policies followed by the Nixon and Ford Administrations concentrated on encouraging improvements in European forces, promoting efficiencies in alliance defence cooperation, and continuing to improve US forces committed to NATO. These policies were intensified under the Carter Administration.

The Carter presidency

The Carter Administation took office with an apparent determination to rejuvenate the alliance. The objective had much to commend it both objectively and subjectively. It was a good opportunity for a new president to reassert American leadership, somewhat tarnished after Vietnam and Watergate, and also for a relatively unknown president to establish his foreign policy credentials. As an indication of its intentions, the Administration appointed Ambassador Robert Komer as Special Assistant for NATO Affairs to the Secretary of Defense. Komer had a thorough knowledge of NATO affairs and was a man with a reputation for getting things done. Within the DoD he became effectively 'Mr NATO', ensuring that alliance interests were always given due consideration in the development of American defence policy. Komer's ideas provided the basis for the two alliance initiatives submitted by the Carter Administration—the Long-Term Defence Programme (LTDP)

The historical background

and the 3 per cent commitment.

The LTDP initiative was proposed by President Carter to a NATO summit meeting in May 1977. The purpose of the LTDP was to assist NATO countries to meet the changing needs of the 1980s, and to counter the trends in the military balance with the Warsaw Pact which were perceived to be running against NATO. It called for a concentrated and coordinated effort in several areas which had been identified as needing urgent attention, and also for a more vigorous pursuit of the means for obtaining, through greater cooperation, the best return on the national resources available. It was particularly designed to incorporate a long-term approach in the hope that national plans and programmes would, in the future, take greater account of the collective needs of the alliance. The areas designated for coordinated action were: readiness, reinforcement, reserves mobilization, maritime posture, air defence, communications, command and control, electronic warfare, logistics, rationalization and theatre nuclear forces.

Parallel to this programme was the proposal that all member countries increase their defence expenditures by 3 per cent in real terms. The exact nature of the pledge was left to be worked out by the defence ministers of the thirteen NATO countries which participate in the integrated command structure. The ministers produced the required guidance on this and other aspects of the programme on 18 May 1977. The guidance specified in part that 'This annual increase should be in the region of 3 per cent, recognizing that for some individual countries economic circumstances will affect what can be achieved: present force contributions may justify a higher level of increase.' A further summit-level NATO meeting, held in Washington in May 1978, approved the programme of defence improvements that had been developed over the preceding year including the 3 per cent commitment. By 1979, the alliance had agreed to sustain defence expenditure and to improve the capabilities through concentration on specific areas.

It is now evident that while the Carter initiatives aimed to achieve greater efficiency in the alliance, the primary objective was to get the allies to do more. Speaking to a closed session of a Senate Committee, Ambassador Komer stated that the rationale behind the LTDP and other initiatives was 'our perception that we needed some kind of an agreed programme whereby we could get the Allies to come along with

The historical background

us in the rearmament we saw as necessary in NATO.' He noted that the US 1978 and 1983 five-year defence programmes already included the bulk of the measures called for under the LTDP. The only countries required to do more under the LTDP than they were originally planning were the European allies. In this respect Ambassador Komer has referred to himself as 'the DoD's chief architect and often negotiator of many ingenious schemes for getting our Allies to contribute more'.[13] And, in the same article, he criticized the Reagan Administration for letting the allies 'prematurely off the 3 per cent real growth hook, thus compromising one of our most promising levers for generating greater Allied efforts'. This emphasis on achieving greater European contributions was somewhat ironic in view of the fact that during the 1970s the Europeans had consistently been improving their defence contributions by an average of approximately 3 per cent.[14] The 3 per cent benchmark was selected precisely because that was what the Europeans had been achieving at that time. It was the US defence contribution that had been on the decline.

The Carter initiatives also re-emphasized the enduring problems facing NATO relating to structure and attitude. The idea behind a special programme was to establish the LTDP outside the normal NATO force-planning process in the belief that only a programme that established goals independent of this normal process would result in actual increases in capabilities. The intention was that NATO nations should agree to undertake programmes in addition to those already planned in their national plans. Paradoxically, however, it was because the LTDP was superimposed on existing plans[15] that it failed to achieve the desired results. Countries fulfilled their commitments where these could be accommodated within existing budget ceilings, but, despite the urgings of the Carter Administration, European governments were unwilling to undertake commitments that had not already been foreseen by their national force-planning cycles.

Despite these problems, the 3 per cent commitment and the LTDP were regarded as the benchmark against which NATO's defence effort would be assessed. However, the Soviet invasion of Afghanistan appeared to make these criteria somewhat redundant and suggested the need for a dramatic change in the dimension and thrust of alliance security policy.

3 The Soviet invasion of Afghanistan

The new strategic challenge

In many senses the Soviet invasion of Afghanistan represented a decisive point in the history of the alliance, since it marked the beginning of a new assertiveness in American foreign policy and, in consequence, the beginning of a serious rift in the alliance. Whereas previously the Europeans had been suspected of not pulling their weight, the months following the Soviet action appeared to provide tangible proof.

For most Americans, the events of late 1979—the seizure of the hostages in Tehran and the invasion of Afghanistan—symbolized the decline of the United States and the failure of past policies based on accommodation and negotiation. From this period on, the United States began to move on a new track towards a national revival, a trend intensified under President Reagan. Unfortunately, this mood did not carry across the Atlantic, since the NATO allies did not suffer the same sense of outrage and frustration, and were reluctant to follow policy prescriptions that derived from an analysis and a public mood which they did not share. If this period did not mark a parting of the ways, it certainly laid the foundations for the tension and recrimination that currently affect the alliance.

The invasion of Afghanistan was interpreted by many Americans as confirmation of a long-held suspicion that the steady growth in Soviet military strength would eventually encourage greater adventurism and expansionism on the part of the Soviet leadership. The use of Soviet troops outside the Soviet sphere of influence was said to represent a decisive break in the unwritten code of conduct governing East–West relations and was an ominous indication of Soviet confidence in the military power that had accumulated over the past two decades. According to this view, the Soviet move into Afghanistan represented an expansion of the Soviet empire at a time of perceived Western weakness and decline, and constituted a serious strategic threat to Western interests, specifically Western dependence on oil from the Gulf region.

For these officials and observers, the Soviet military intervention was a much-needed catalyst to awaken the Western alliance to the true nature and scale of the Soviet threat. Accordingly, they called for a coherent and vigorous response from the West. Describing the Soviet action as the most serious crisis since World War II, President Carter initiated a two-pronged response. First he introduced a number of essentially short-term punitive measures intended to demonstrate to the Soviet Union that aggression carried a high political and economic price and, second, in stressing the strategic consequences for the West, he called for a coordinated and collective response by the Western allies and Japan to this new strategic challenge.

There can be little doubt that the Soviet action was a personal blow to the hopes and objectives of President Carter, and that equally it confirmed the suspicions of a number of his advisers and critics alike. However, the US response cannot be assessed by reference to the invasion alone, but must be viewed against the political conditions that prevailed during that period. The perception had been steadily growing in the United States that America's position in the world was on the decline, particularly relative to that of the Soviet Union. A number of developments had contributed to that perception: a constant emphasis on what was termed 'the unprecedented build-up' of Soviet armed forces, and the expansion of Soviet influence world-wide, either direct or through proxies; the specific concern at the strategic nuclear level that American land-based missiles were vulnerable to a Soviet first strike; and the evident dependence of American society on a regular flow of oil from the Middle East. The coincidence of these factors meant that American society began to experience a sense of vulnerability to pressures and events beyond its control, a sensation that was both unfamiliar and unwelcome. The SALT II ratification hearings drew attention to many of these weaknesses, and to the fact that America had, in the words of Senator Javits, 'goofed off' with respect to its military strength.[1]

The perception of American weakness was dramatically symbolized by the seizure of the hostages in Iran: suddenly America's impotence was evident for all to see. The humiliation of this episode was placed in stark relief by the Soviet invasion of Afghanistan. The contrast between these two events, between the Soviet armed forces on the move and American military helplessness, confirmed the prevailing suspicion that the balance of power had swung to the advantage of the Soviet Union. The President's personal standing during this period is also relevant to an understanding of the US response to Soviet actions. Rightly or wrongly, President Carter's image was one of a vacillating and uncertain president, hardly an appropriate image with which to be entering an election year. In view of this reputation and the general feeling of frustration throughout the country, anything less than a robust American reaction to the Soviet invasion would have been political suicide. In addition to his evident personal disillusionment with the Soviet Union, it was crucial that the President should show himself capable of strong and decisive leadership.

The American Administration therefore sallied forth with a programme of action for the alliance designed to counter the new strategic challenge. The assumption underlying the programme was that NATO strategy must be changed in accordance with the new circumstances, and the allies were expected to march unhesitatingly to the new tune emanating from Washington. The stated intention behind the American approach was to develop a shared assessment of the international situation and its implications for Western security, and to agree upon a programme of action for the alliance reflecting a division of effort and a sharing of burdens. While Administration officials stressed the 'shared' nature of its new policies, in fact they were essentially driven by the initial American assessment that the Soviet action represented a new strategic challenge. There was no room for doubt about this assessment, and in this respect officials in Washington appeared to believe that they had a monopoly on strategic wisdom. Hence the allies were informed rather than consulted on the main elements of the alliance response.

A number of European governments, however, were uncomfortable with the American approach. European reactions to the invasion of

The Soviet invasion of Afghanistan

Afghanistan had been more restrained than those of the United States. All allied governments condemned Soviet actions as violating the norms of international law and practice. But most Europeans were unwilling to accept 'the bear on the move' theory that appeared to underlie the American analysis; they preferred a more measured assessment of the strategic consequences of the Soviet intervention than the American interpretation allowed. Also, they regarded the security and stability of the Gulf region as lying more in the implementation of political and economic measures than in the military efforts that the United States was proposing. There were fears that hastily contrived regional defence arrangements would only contribute to the very instability that they were meant to avert. There was general agreement within the alliance that the Soviet occupation of Afghanistan and the instability in the Gulf were matters of direct concern to the West and required careful evaluation. But few governments wished to go as far or as fast as the Administration was proposing. The Europeans' reticence was also connected with the manner and haste with which the American approach had been conceived. To many Europeans it appeared that the actions of the Carter Administration had been dictated as much by the requirements of an election year as by a dramatically changed strategic situation.

The Europeans were in a difficult position. The political climate in Washington had become distinctly critical, and in many cases openly hostile, towards the European position. The dominant perception was that America's allies had been inexcusably reluctant to support the United States, first during a time of great American need, the seizure of the hostages, and subsequently in response to a situation in which their own interests were at stake, the Soviet invasion of Afghanistan. Although the two crises were quite different in origin, they shared certain fundamental consequences, notably in terms of maintaining stability and excluding Soviet influence in the Gulf region, but also in terms of ensuring alliance cooperation and solidarity. Most significant, however, they had become blurred as regards the American public's perception of European actions—that, in a time of crisis and need, America's allies had put self-interest above alliance solidarity. In this politically sensitive climate, European governments were left with little alternative but to endorse the policies emanating from Washington. To have done otherwise would have risked serious rupture within the alliance. Alliance politics demanded that, in the short term at least, the Europeans supported both the American analysis and the proposals that evolved from it.

The division of labour

The Soviet invasion of Afghanistan and the American call for a collective Western response focused attention once more on the perennial question of alliance defence effort and burden-sharing. American concern had already begun to register over the European performance, or lack of it, in implementing the LTDP and in fulfilling the 3 per cent commitment. Now there was a new threat to the alliance, but one that lay outside the traditional boundaries. Simple logic suggested that if a new dimension had been added to Western security, then this would involve new responsibilities, which would in turn necessitate the provision of additional resources. However, simple logic did not have to account for the NATO force-planning process, or the economic circumstances of the time, or the different views of what the alliance should do.

The Carter Administration suggested three general areas in which additional effort was required: an increased military presence in the Gulf region in peacetime, both on a permanent and on a temporary basis, in order to demonstrate the Western commitment to defend the region; a credible capability to introduce significant combat forces into the region during a crisis or conflict conditions; and the need to provide security and economic assistance to the regional states of the Gulf. American officials emphasized that the resources for these additional areas must not in any way affect the credibility of NATO's defence posture in the Central Region.

The American initiative concerning the security of the Gulf placed the Europeans in an awkward predicament. There was general agreement that events in the Gulf were relevant to Western security, but there was no real consensus over what the alliance should do about it. Many Europeans were uncomfortable with both the American diagnosis and the prescription. European reluctance concerning the new policies was also related to capabilities and resources. Few Europeans had the means to help the United States directly in the Gulf, which meant that, having agreed to the policy, they would have little influence over it, and none was in any position to find the additional resources that the United States was suggesting the new situation demanded.

The determined advocacy of the Administration and rumblings from Congress about American boys dying for European oil were sufficient to propel the allies, despite their misgivings, down the path prepared by the United States. Alliance politics dictated that the Europeans should subscribe to policies that they were neither sure of nor capable of funding. Salvation appeared through the resurrection of the 'division of labour'. It was agreed that since the United States was best equipped to provide for the security of the Gulf, the allies would consequently have to assume a greater burden in Europe. In classic alliance fashion, the differences in attitude and capabilities between Europe and the United States concerning an appropriate alliance response were effectively reconciled. Interviewed in *Die Zeit* (1 Feb. 1980), the Federal Republic's Foreign Minister, Hans-Dietrich Genscher, explained:

> There are common interests in this matter, but not all members are equally fit for the same task. There are members that because of their tradition, history and engagement are more fit than the FRG to safeguard the outer framework conditions... It is much rather a joint and coordinated strategy that is involved, where every partner will take on the task he is best suited for. Division of labour is necessary.

The Europeans agreed that they would give what support they could to the United States in the Gulf, but for the most part would concentrate on their tasks in the European theatre. If the new American commitments created some slack in Europe, this would be taken up by the Europeans. However, there was a drawback to these promises of intent: new tasks would be taken up only if they did not entail more expenditure, and only so long as they did not jeopardize existing ones. In practice, the European share of the new division of labour amounted to little more than a reshuffling of existing priorities; no new commitments or new resources were added.

The US role

Under the Carter initiatives, the United States agreed that it would take prime responsibility for establishing a military presence in the Gulf. This would take the form of deployment of naval forces, the formation of rapid deployment forces, and the securing of base facilities in the region.

The concept of a Rapid Deployment Force (RDF) had existed well before 1979, but was given a new urgency by developments in south-west Asia. However, considerable confusion existed both in the US and abroad about the precise objectives, and therefore force structure, of an RDF. What sort of force should it comprise, and what logistic components would it require? Department of Defense officials stated that the number and composition of an RDF would vary with the nature and location of the crisis, and would be drawn from existing forces. Even if this proposition were accepted, it was evident that tremendous problems remained in terms of transporting and supporting such a force.[2]

The new policy of providing a military intervention capability for the Gulf region meant a search for local facilities which would support such operations. Agreements were signed with Oman, Kenya and Somalia for access to a number of bases and facilities in those countries. In general, regional states proved extremely cautious in the provision of these facilities, preferring US forces to be 'over the horizon' rather than 'on the horizon'. Egypt has proved to be the most cooperative state in this respect, even permitting US forces to carry out deployment exercises on its bases.

It was evident that the United States would face major problems in implementing its decision to create a deterrent capability in the Gulf region. One of the key questions for the alliance was the degree to which this new commitment to the Gulf would affect America's existing commitments to NATO. It soon became clear that it would have substantial consequences: circumstances could arise in which US resources, both men and transport, that would have gone to Europe would now go to south-west Asia. Simultaneous crises in Europe and south-west Asia would produce a shortfall in American capabilities. In this respect, attention turned fairly naturally to the allies. US defence officials emphasized that if Japan and Western Europe were able to assume a greater security burden, then the United States would be in a much better position to respond to Soviet moves in other areas.

The Soviet invasion of Afghanistan

The European contribution

The contribution that the United States sought from the allies fell into two categories: direct assistance and indirect assistance. Direct assistance was sought through the deployment of armed forces, particularly naval, the intensification of economic assistance and political influence, and permitting the United States the use of local facilities to ease American deployments. While a degree of coordination could be effected through the alliance framework, these actions would be carried out on a bilateral basis independent of the formal NATO command. Indirect assistance would be given by assuming additional tasks in the defence of Western Europe, thus freeing American resources for employment in the Gulf region. These actions could be carried out within the formal NATO framework.

Direct assistance. The request for an allied commitment in terms of actual capabilities referred essentially to naval forces. While several NATO members possessed naval assets that could be sent to the Indian Ocean on an occasional basis, only Britain and France were capable of providing naval forces on a permanent basis. The French Indian Ocean force fluctuates between twelve and eighteen units, depending on circumstances. Periodic reinforcements include one of the two French aircraft carriers.[3] The British have traditionally deployed a Royal Navy task force to and through the region approximately once a year. Any proposal to secure a permanent European contribution to an Indian Ocean naval presence would have to weigh the implication on capabilities in other areas. Such a contribution would inevitably detract from capabilities earmarked for missions elsewhere, notably the Atlantic, where SACLANT has consistently stated that he has insufficient assets to satisfy his requirements.

In addition to potential naval contributions, it should be noted that Britain and France maintain forces that could operate effectively outside the NATO area. Both have marine and airborne units for quick reaction contingencies, but both suffer deficiencies in the logistic capabilities necessary to deploy and sustain these forces. However, the recent British success in deploying and sustaining a substantial force of men and ships to regain the Falklands was a remarkable demonstration of Britain's capabilities in this respect.

The American proposals emphasized the contribution that the allies could make to the stability of the Gulf region through the expansion of existing political influence and the development of military and economic assistance programmes to nations in the region. A network of political and economic ties, including several armament and military training programmes, exists between European nations and the Gulf states. However, while European members are willing to examine methods of ensuring the stability of the region and improving Western influence, they are cautious of becoming too closely involved with regimes that lack popular support.

Despite the differences that exist over the type and degree of alliance involvement required in the Gulf, the area of greatest transatlantic divergence continues to be the question of a Middle East settlement. Many Europeans believed that the American emphasis on the military threat to the Gulf missed, and even distracted from, the real basis of instability in the region—the failure to produce a solution to the Arab-Israeli dispute. The Venice declaration of the Nine in June 1980, and the European insistence that alternatives to the Camp David process must be sought, were both sources of irritation for the United States. However, the latest Reagan initiative and the tragic developments in the Lebanon have, for the moment, dampened the differences between the two sides. Furthermore, the participation of British, French, Dutch and Italian units in the Sinai peace-keeping force, and of French and Italian units in the Lebanon peace-keeping force, has provided tangible proof of European willingness to assume responsibility and to be actively involved in the maintenance of stability in the region.

Finally, the United States has indicated that those European partners located on key ocean routes to south-west Asia could make an important contribution to a Western presence in the Gulf by providing the United States and other allies with the use of *en route* base access. Yet this apparently cost-free proposal demonstrates perfectly the inherent problems of correlating an overall strategic assessment with the attitudes and interests of individual nations. The Portuguese, Greek, Turkish and Spanish authorities have all, at various times, indicated a reluctance to allow their American bases to be used for Middle East contingencies.[4]

Indirect assistance. Discussion of a European role in the new strategic situation has concentrated on the proposal that the allies take on more tasks and responsibilities in the Central Region. This would consist of compensating for the eventuality that American forces previously earmarked for the European theatre would be used elsewhere. The United States has, in recent years, placed heavy emphasis on improving the readiness and reinforcement aspects of its European commitment. This has involved the prepositioning of equipment for two divisions and one armoured cavalry regiment and includes plans for equipment for three more divisions by the end of 1982. Under current circumstances, it is highly unlikely that these 'in place' American troops or their immediate reinforcements would be affected by the Persian Gulf commitments. The most likely contingency is that American reinforcements earmarked to move to Europe at a later stage, and the transport needed for their movement, would be assigned for use elsewhere. It is the absence of these resources for which the Europeans must specifically be prepared to compensate.

What precisely this compensation would mean in terms of additional capabilities, and what additional resources would be needed, has not been made explicit. In order to demonstrate an alliance response to the new situation, the NATO military authorities selected a number of measures already contained in the LTDP which they said should receive higher priority, and thus earlier attention, by nations than already planned. These measures fell into two categories: a first phase involving war reserve stocks, electronic warfare, and defence against chemical weapons; and a second phase involving the activation of additional reserve units, maritime forces, readiness, and the wartime commitment of civil aircraft and merchant ships.

General agreement was reached that member countries would attempt to absorb these new priorities into their defence plans. But without the provision of extra funds, the acceptance of a new priority would inevitably result in the downgrading of a planned or existing capability. In this sense the attempt by NATO to coordinate a special response to developments in the Gulf ran into the all too familiar problem of national force-planning and resource limitations. Recommendations for special action were followed only insomuch as they could be reconciled with existing national priorities and plans. If something additional was done in one area, then something had to drop out of another.

4 The alliance dilemma

Despite constant pressure from the United States, including the LTDP, the 3 per cent target and the new strategic concept to deal with regional threats, the dilemma facing the alliance at the end of the Carter presidency was that European governments had not produced the additional defence effort—either in resources or in responsibilities—as suggested by the American analysis of Western security requirements. This situation was quite familiar, but the dilemma was made more serious than usual by the prevailing political climate. American public and Congressional interest had become engaged in alliance affairs, and the gap between what the allies had promised and what they had delivered was all too visible.

Implementation of the much-vaunted LTDP had been disappointing. Because it was superimposed on the normal planning cycle, countries had been for the most part unable or unwilling to absorb its recommendations unless they could be accommodated within existing programmes. This is not to say that the LTDP achieved nothing. The fact that alliance members collectively considered alliance requirements for the next fifteen years in a number of key areas was itself an achievement. Furthermore, several important measures were accelerated as a result of LTDP recommendations. But as the LTDP was not accompanied by additional resources, its impact was inevitably limited. Consequently it became a vehicle for demonstrating the failure of the NATO allies to follow through on their commitments. General Rogers, the current SACEUR, reflected this critical approach in his testimony in March 1981 to the House Armed Services Committee: 'Too many security commitments have become overdue promissory notes: we have

The alliance dilemma

slippages, reductions and cancellations in nearly every Allied nation, including the United States.'

In a similar fashion, the 3 per cent commitment also provided a convenient target for criticism. The original idea of the 3 per cent had been to provide a symbolic expression of NATO's collective determination to sustain a credible defence. However, owing to the deterioration of economic conditions, most nations found the commitment difficult to meet. Instead of serving as a symbol of alliance cohesion, the 3 per cent became the perfect instrument with which American critics could berate the allies. Furthermore, rather than providing a means of ensuring that nations improved their defence capabilities, the 3 per cent became the subject of what David Greenwood has termed 'creative book-keeping'.[1] As Greenwood has pointed out, the 3 per cent proved harmful in two respects. First, it suggested erroneously that the provision of extra funds equalled the provision of extra capabilities—it took no account of how resources were spent, whether on pensions or on weapons. Second, it became the focus of public interest and thus distracted attention from the real problems of improving NATO's defence capabilities.[2]

Alliance shortcomings as regards the LTDP and 3 per cent commitments were largely explained by the increasingly severe economic restraints under which defence budgets had to operate. The prevailing economic conditions in most European countries, involving a decline in economic growth, pressure for cuts in government expenditure, and consequently severe competition for the allocation of increasingly scarce resources, meant that no government was willing to allocate additional funds for defence. In addition to these general economic conditions, defence expenditure suffered from a number of factors peculiar to the defence budget, notably the higher inflation rate in the defence sector. The rising costs of personnel, of increasingly sophisticated equipment and of fuel meant that even when a 3 per cent real increase was achieved, it was only barely sufficient to keep a country marking time: it would meet existing and planning commitments but there was no excess to purchase additional capabilities.

It was under these circumstances that, during 1980, NATO ministers had endorsed the US proposals for the extension of alliance responsibilities to the Gulf. However, while they agreed to the post-Afghanistan measures, all ministers argued that additional commitments could be carried out only within the existing framework of the 3 per cent annual increase.[3] This was hardly the response that the United States and the NATO military authorities wanted. General Rogers expressed the impracticability of this position: 'There was an uncomfortable tendency on the part of some nations to suggest that such defence improvements should somehow be made without the allocation of additional resources, a suggestion which I believe to be unfeasible.' And he emphasized that, in order to meet the challenges outside the alliance, NATO must provide additional forces and resources and 'not simply redistribute existing ones'.

Despite such admonitions, it was evident that additional expenditure from the European allies would not be forthcoming. 'Doing more' would be achieved only by giving priority to certain tasks or measures at the expense of planned or existing capabilities. Improvements in certain areas or the adoption of new tasks would result in the postponement or elimination of other capabilities. Inevitably, 'doing more' would necessitate a search for ways to do better with existing resources. This would involve a renewed emphasis on increased cooperation in defence procurement, a more effective specialization of tasks, and the utilization of new technology. But such ideas were hardly novel; they had in varying degrees been the focus of alliance attention for many years, and had failed to yield any dramatic improvements. They would not produce an easy way out of the limited resources predicament, nor would they conceal the basic fact that Europeans were unwilling to spend more on defence.

The formula adopted by the alliance in response to the American initiatives was perfectly consistent with established alliance practices, and the traditional fudging of issues in order to conceal disagreement. On past experience, it was perfectly conceivable that the adoption of the 'division of labour' concept would reconcile the evident differences of approach and effectively camouflage from public opinion the discrepancies between what the United States wanted and what the Europeans were willing to contribute. The alliance would emerge slightly battered, but unscathed, from this crisis, as it had from others. However, the optimism behind this prediction took no account of the political climate. As already discussed, the early months of 1980 had seen a substantial change in the attitude of American public opinion, as represented by Congress and the media, against the European allies. This attitude did not diminish; if anything, it intensified. Throughout 1980, the approach in the media was consistently critical—the allies were weaseling out of the 3 per cent commitment; the United States could no longer afford to bear the burden of Western defence alone; the United States should no longer subsidize the Europeans, unless the allies were willing to do more on their own behalf; the United States should reduce its commitment.

The Europeans replied to these criticisms by using the all too familiar arguments that it was not what was spent that counted, but what was achieved in terms of manpower, equipment and overall readiness, and that in this respect they were doing their fair share. They pointed out that European forces provided the major proportion of NATO's immediately available forces. The Federal Republic in particular emphasized the substantial contribution of the Bundeswehr and the large number of reserves made available through conscription. However, these arguments fell on stony ground. The perception of 'free riding' was too deeply entrenched in the American psyche to be mollified by such evidently self-serving explanations, particularly when European shortcomings were so clearly visible.

American public concern over these issues and over the broader question of the continuing relevance of the NATO Alliance did not disappear. It thus remained to be seen how the new Administration of President Reagan would reconcile the obvious differences in approach on either side of the Atlantic concerning the requirements of Western security.

5 The Reagan Administration

The most striking feature of the new Administration was its ideological conviction. The Reagan team took office with very clear ideas about what was wrong with the United States and what they were going to do about it. For the past decade, they had watched helplessly as, in their view, American power and influence steadily weakened while the Communist menace continued to grow unchecked. Now, like prophets from the wilderness, they approached the task of reversing these trends with an evangelical zeal which initially left little room for listening to the views of those who did not see the world quite as they did. Some officials appeared to believe that NATO's problems, and current European attitudes in particular, derived from the allies' perception of America's decline and their loss of confidence in American leadership. According to this view, the future of the alliance depended on a reassertion of American leadership based on a renewal of American military strength: in order to reassure uncertain allies America must lead by example.

Several Reagan officials had long criticized the shortcomings of the NATO allies, and their tolerance level for what were seen as European 'special pleadings' was, to say the least, low. Their attitude amounted to 'shape up or else'. This tendency to look critically at the alliance was accompanied by a disposition within certain areas of the Administration to look beyond NATO for the central focus of American foreign policy. The California background of the President and several of his Administration provided a natural tilt towards the Pacific region as being at least as important as Europe to American security—and, in view of developments in Europe, perhaps even more so.

The Reagan Administration

These aspects were hardly conducive to the mutual tolerance and understanding that were essential if outstanding differences in the alliance were to be settled. The situation was made worse by the coincidence of a number of disputes—some directly related to the alliance security relationship, others not—which all served to fuel the tension already existing on either side of the Atlantic. Despite the profusion of disputes, the most significant issue for the long-term future of the alliance remained the question of Western security requirements. The inflexible attitude adopted by the Reagan Administration on what needed to be done ensured that this issue would remain an area of considerable contention.

The Reagan defence budget

President Reagan took office with a clear commitment to improve America's defences. He and his officials constantly reiterated their belief that an obsessive preoccupation with the policies of détente by previous Administrations had seriously weakened America's defences. By contrast, the Soviet Union, during the same period, had engaged on what was termed 'an unprecedented military build-up'. As a result, in the President's view, the military balance had swung against the United States—a situation which had ominous implications for the Free World. In order to reverse this situation, the President determined that American defence spending should be dramatically increased.[1]

His 1983 request was the largest peacetime increase in defence spending in US history. Even allowing for the difference between authorization and appropriation in the American budgetary cycle, the increases will be significant. The emphasis on defence is all the more remarkable because it runs counter to the central feature of the President's economic policy. In his efforts to reduce the budget while at the same time reducing taxes, the President has insisted on major reductions in all areas of government spending, many of which will have serious social effects. Defence spending, however, has been set apart from other areas of Federal expenditure. In the current situation, which is frequently described in terms akin to those of a national emergency, defence is sacrosanct and sacrifices necessary: 'We must now pay the bill for our collective failure to preserve an adequate balance of strength

The Reagan Administration

during the past decade or two. While our principal adversaries engaged in the greatest build-up of military power ever seen in modern times, our own investment in forces and weapons continued to decline until very recently.'[2]

However, there are signs that the public support for more spending on defence that was present in 1980 has begun to evaporate. Many believe that economic reality will force the President to realize that he cannot pursue higher defence spending at the rates projected and at the same time reduce the budget deficit. Moreover, an increasing number of Congressional voices are now suggesting that, at a time of economic hardship, defence can no longer be given privileged treatment.[3] Yet, despite the rising tide of opposition, the President has shown himself to be unwilling to back away from his commitment to higher defence spending, and has required only small cuts in the previous programme. Whichever direction American defence spending takes will be of direct significance to the alliance in the sense that higher defence spending means pressure for the allies to reciprocate, whereas a search for reductions will inevitably turn to the NATO commitment to see if savings can be made. For the moment, however, it is the capabilities that the United States plans to purchase with its prospective increases, rather than the increase itself, that are of greatest interest to the alliance.

In its allocation of resources, the Reagan Administration has accelerated the trend begun under President Carter of preparing to respond to crises outside the NATO boundaries, mainly in south-west Asia. However, President Reagan has gone considerably beyond the Carter strategy by proposing that the United States adopt a global approach to American security. This global approach rejected previous assumptions of one and a half or two and a half war scenarios as the basis for military planning, and calls for the development of capabilities to meet a wide range of global contingencies. The United States must be ready to meet Soviet aggression wherever it occurs, but it must also be able to take counter-offensive action at points of its own choosing, not necessarily on the aggressor's immediate front, but at his more vulnerable points. This concept has been termed 'horizontal escalation'.[4]

The shift towards this global strategy has produced several fundamental changes in the planning of American ground forces, notably: a reallocation of substantial portions of the ground and land-based air forces from their previous reinforcing role to Europe to the Rapid Deployment Force; an acceleration and expansion of the strategic lift programmes for moving forces to south-west Asia; and a substantial increase in naval power, particularly the offensive striking power of the fleet, in order 'to threaten the Soviets and their surrogates with the prospect of a wider war if they seem to be getting the upper hand in a vital area like South-West Asia'.[5]

Administration officials have articulated the objectives of current American policy towards the Gulf as being to prevent Soviet domination over the Straits of Hormuz and the shipment of oil. Noting that the Soviet Union would probably be an energy-importing country in a few years' time, Secretary Weinberger stated that this would certainly increase the possibilities of the Soviets trying some adventurous tactics into the oilfields. Enough of an American presence was therefore required in the area 'to lead the Soviets to conclude correctly that there would be an unacceptable risk to them to try the sort of thing they are trying in Afghanistan'. Outlining American defence planning for the Gulf region, former Deputy Secretary of Defense Carlucci argued that since substantial advance warning would be received of any large-scale Soviet aggression in the Gulf region, and since the distance is great and the terrain difficult, a combination of American air power and RDF land forces could confront a Soviet attack. This would assure the Soviet leadership that the time involved in such a war would be protracted and force them to calculate the course and consequences of a long war: 'In sum, the RDF should deter the Soviets from executing a lightning-like thrust into the Gulf region. Deprived of that option, any Soviet design for aggression in the Gulf region must anticipate a longer war and the possibility of a United States flexible response in other areas.'[6]

The RDF has been substantially expanded both as a concept and in the capabilities that it will have at its disposal. Its area of operations will cover twenty countries (but not Israel), and it will have authority and responsibility for US military activity in the region of the Persian Gulf and south-west Asia including such responsibilities as providing security assistance. The headquarters at Tampa, Florida, was scheduled to become a full-scale military command in January 1983, with over 800 personnel from the army, navy, marines and air force. To undertake its mission, the Reagan Administration has ordered the army to assign five divisions (compared with three now), the marine corps two divisions and their air wings (compared with one and a half now), and the air force ten tactical fighter wings (compared with five now). Naval forces will include three aircraft carriers and escorts, and thirteen ships with weapons and supplies based at Diego Garcia.

In an attempt to give substance to the term 'rapid', improvements have been carried out on the C141 and C5 aircraft. With regard to the rapidity of moving the RDF, the commander of the deployment force commented that, with adequate warning, a tactical fighter unit could be on the scene within several hours, an amphibious battalion of 1,800 marines could be ashore within 48 hours, and an army airborne brigade of 3,000 paratroopers within four days.[7] The remaining two-thirds of the combat elements of an army division could be there within two weeks. However, it would take 35 days to bring in supporting heavy artillery and a mechanized division with its tanks. Reinforcements and replenishment of supplies would take even longer despite the recent acquisition of high-speed ships.

After Congressional criticism that objectives established by Secretary Weinberger were widely ambitious and lacked serious strategic analysis, the Administration ordered a Defense Guidance in order to establish priorities for the armed forces for the period 1984-8. President Reagan's National Security Advisor, William P. Clark, provided the basic outline of this Guidance in a public policy speech, in which the defence of North America, NATO and south-west Asia were given as the three main priorities.[8] The Guidance appeared to modify the Weinberger objectives by establishing priorities for general planning; in terms of priority for resource allocation, however, it leaves the situation somewhat blurred by calling for a dual emphasis on forces for the Gulf and for NATO. Arguing that the two are strategically connected, it calls for stretching forces between them: that is, providing forces that would be used both in the direct defence of NATO and in the defence of allied interests in south-west Asia.

Despite this attempt by the Guidance to establish priorities in American defence planning, critics still insist that America's objectives lie beyond its resources. They argue that even under the Reagan spending proposals, the United States cannot afford to do all the things it wants to.⁹ The Reagan Administration, it would appear, has encountered the problem of matching over-expanding requirements with limited resources. However, any satisfaction that European governments may have from watching the American government come face to face with the realities of defence spending will be shortlived, since the result of this encounter will certainly have a number of consequences for the alliance, most of them negative.

In terms of resource allocation, it is evident that the south-west Asia commitment will eventually compete with NATO. Department of Defense officials insist that the two regions are equally important. Instead of one spotlight centred on NATO as the focus of American defence policy, there are now two, the other being on south-west Asia. As a result, the first may shine with less brilliance. Many observers fear that, given the levels of resources in terms of the men, equipment and logistics required to make the RDF credible, there will inevitably be a conflict with NATO commitments for resources. A senior Senate staff member commented that the threat to America's commitment to NATO was not on Capitol Hill in the form of Mansfield-type amendments; the 'more sinister long-term threat lies in the Department of Defense itself'.¹⁰ In his view, the Department had been asked to work to an impossible set of objectives, which would inevitably produce 'a constant state of competition between those tasked with making the RDF work and those working in NATO'. Everybody would be competing for very limited mobility assets and scarce defence dollars; it would be a choice between extra divisional kits for POMCUS (prepositioned stocks) in NATO and war reserve kits for the Gulf. In this situation, given the current state of antipathy in certain areas of the Department towards the allies, 'NATO could easily get burned'. It was already possible to detect the view that America's real priority was the protection of Gulf oil, and that the NATO commitment represented an excessive burden. 'You can see a slide away in terms of banking up the declared commitment until you get to the point where you say we are pouring so many dollars down the hole to keep 350,000 gaps in Europe; there's not enough to make the RDF work, get some of that leading edge and the tail required to back it up out of Europe and reallocate both to the Persian Gulf scenario.'

Such predictions may be unnecessarily alarmist, but the view that

The Reagan Administration

resource constraints will force the United States to make some fundamental choices in its defence planning has been widely echoed. The attempts by the Reagan Administration to 'fight on every front' has provoked a debate within the US defence and foreign policy community as to what America's priorities should be. Many believe that the Administration should take the assumption implicit in its current emphasis on a global strategy to its logical extension: that is, the United States should stop trying to contest Soviet military predominance on the Eurasian land mass and concentrate on exploiting America's maritime superiority. This approach, known as 'global unilateralism', would involve a progressive detachment from Europe.

Proposals to reduce the American commitment to Europe have been put forward both by those in favour of higher defence spending and by those who seek reductions. The theme common to both groups is that American defence policy should be based on priorities more appropriate to America's capabilities and real interests. Those who opposed the excessively high levels of military expenditure involved in the Reagan defence budget question the need for the continuing military investment in Europe on the grounds that the Europeans themselves have not seen fit to reciprocate. They argue that since the bulk of any defence budget goes on general purpose forces, only this area will yield significant savings, and this necessitates looking at NATO: 'To be serious about cutting defence spending, you must talk about America's major alliances—particularly NATO, which is costing us half our entire defence budget.'[11] The most familiar argument in favour of troop withdrawals from Europe is that if Europe cannot be bothered to defend itself, then America should stop trying to do so on its own. Those supporting withdrawals recognize that a strong and independent Europe is vital to the political and economic well-being of the United States, but many argue that American reductions are now necessary in order to jolt the Europeans into making a far stronger effort to defend themselves. Some go even further, alleging that the failure of the NATO allies to bear their fair share of the military burden jeopardizes not only the security of Europe but also the physical safety of US forces stationed there: 'US troops are now in serious danger because our European allies refuse to back them with adequate defence forces. We shouldn't allow them to become cannon fodder in some future European war.'[12]

The Reagan Administration

Implications for the alliance

Whatever course the debate over America's defence priorities takes, it is clear that pressure on the allies for greater defence contributions will continue unabated. The Administration has, from the beginning, expressed the view that the allies do not do enough: 'We note that our allies are, in many cases, not contributing their equitable share of the defence burden in their own theatre, despite having the obvious ability to do much more.'[13] And similarly: 'It is clear that to achieve greater equity among the burdens imposed on the salaries and taxpayers of each nation and greater safety for us all, several of our allies will have to assume a larger share.'[14] Despite the fact that the Special Report to Congress on alliance burden-sharing reported favourably on the overall performance of the NATO allies, DoD officials still argue that the burden is being borne disproportionately: 'When all is said and done, the richer are doing less than the poorer countries and the average American is subsidizing the defence of his wealthier European counterpart.'[15]

One of the key assumptions underlying the new global strategy is precisely that the United States has rich allies who can afford to do more. The existence of these allies means that the United States can devote its attention to the West's greatest deficiencies—the threat to south-west Asia, and the inability of the US navy to carry out certain naval missions, particularly offensive operations against Soviet forces and territory. Secretary of Defense Weinberger has consistently emphasized this point: 'These general perceptions have guided our allocation of general purpose forces and funds somewhat away from the "traditional" theatres of Europe and North-East Asia, where relative threats are less and we have rich allies who can do more on their own behalf, and towards the defence of South-West Asia.'[16]

The Reagan Administration has produced detailed proposals of the measures that they believe the alliance should be carrying out within the division of labour, particularly in support of American plans for south-west Asia. Alliance support has been defined in two specific areas, *en route* access and combat service support in Europe, to compensate for the reinforcements diverted from the NATO contingency.

En route access remains a politically sensitive issue, with few countries being willing to make the type of advance commitment that the United States would like. However, DoD officials have expressed their satisfaction with the cooperation provided by their allies in the way of *en route* access during Operation Bright Star in October–November 1981. A number of NATO members, including the United Kingdom, Portugal, Spain, Italy, Germany and Turkey, provided support facilities for the deployment of US-based airborne troops to Egypt, Sudan, Somalia and Oman.

With regard to combat service support, American officials believe that in a crisis, because of the lack of infrastructure in south-west Asia compared with the European theatre, up to 60,000–70,000 US-based combat support personnel who would have gone to Europe would be sent to the Gulf. Such a redeployment would obviously place an additional responsibility on the host nation support programme in Europe. According to US officials, the reaction of the NATO allies to these detailed requests has been non-committal, and in traditional fashion the NATO authorities are currently studying their implications.

American pressure on the allies to increase their defence expenditure has continued under the Reagan Administration, even to the extent of retaining the 3 per cent target figure as an indicator of comparative defence effort. Early pronouncements by US officials suggested that the Reagan Administration was prepared to move away from fixed percentage increases as a means of measuring defence effort. However, despite these indications, and the opposition of several NATO allies notwithstanding, the United States insisted that the Final Communiqué of the May 1981 NATO ministerial meeting should record confirmation of 'the standing Allied commitment to the 3 per cent formula guidance' in national military budgets.

If, however, the pressure on the allies to spend more is consistent with past practice, the motivation behind this pressure is somewhat different. This was illustrated by former Deputy Secretary Carlucci: 'We want to be able to say a new awareness has risen in the Alliance, a new consensus to give first priority to the defence of freedom... we want to demonstrate that our allies and friends are contributing their fair share of the common burden.'[17] And similarly: 'The American people will not want to march alone. If our effort is not joined by all who are threatened, by all who face the common danger, we in the United States could lose at home the critical public support for which we have laboured so long and so hard.'[18]

The Reagan Administration

The essential realization that these expectations will not be fulfilled, and that European governments are not willing to impose the type of sacrifices being demanded by the Reagan Administration, will almost certainly sustain what is a low but perceptible momentum against the alliance. Antipathy, rather than hostility, is an appropriate expression for the prevailing sentiment in Washington — antipathy which arises from the different perspective and orientation of the Reagan Administration and from the growing feeling that America's interests lie outside the traditional European focus. But there is also frustration among a number of officials and Congressmen at the persistent failure of the Europeans to react to the self-evident dangers threatening them.

It is also recognized that, with the exception of the State Department, the Reagan Administration is less willing than its predecessors to defend the interests of the alliance against Congressional criticism. European governments regretted the resignation of Secretary of State Haig precisely because he was believed to be the sole defender of European interests. Senate staffers have noted that the Department of Defense has been frequently less than alert concerning the passage of legislation which could adversely affect alliance interests.[19] There is no figure playing a role comparable to that of Ambassador Robert Komer in the Carter Administration in defending alliance interests on Capitol Hill. In his recent report, Senator Nunn commented on this vacuum: 'In our own Department of Defense, there is today no high-level person working full-time on these important Alliance cooperation issues. It is not until you reach down in the bureaucracy to the Deputy Assistant Secretary of Defense for NATO affairs level that anyone is identified for this focus. This position has been vacant for over a year.'[20]

Career officials in the Department who were working in the field of cooperation with the allies in weapon procurement also noted a change of policy emanating from senior officials.[21] They commented that it was now more difficult to get high-level approval for cooperative projects with the allies, particularly those involving the release of technology to the allies. This reluctance emanates from the general objective of the Reagan Administration to reduce the export of high technology to the East. It is difficult to reconcile with the most recent initiative to exploit emerging technology — an initiative that will obviously require a substantial increase in alliance cooperation.

The Reagan Administration

So far, discussion of the problems currently facing the alliance has been confined to the vexed question of the resources and responsibilities necessary to sustain Western security. Despite the evident differences on either side of the Atlantic about what needs to be done and the different emphasis of this Administration, it is quite conceivable that if dealt with in isolation these issues could be reconciled. Over the years, the alliance framework has proved to be extremely adept at quietly defusing and accommodating such differences of view. Unfortunately, defence-related issues do not evolve in a vacuum, and it is the political environment within which the current debate is taking place that makes its resolution difficult and even doubtful.

The political environment

In assessing the future health of the alliance, two aspects of the current political environment are particularly relevant: first, the concentration and intensity of the issues causing friction between the United States and Europe (differences exist at almost every level of interchange — economic, political and military); and, second, the consequent increase in public frustration on both sides of the Atlantic concerning the costs and benefits of alliance membership.

In Europe, public discontent with the alliance has centred on the role of nuclear weapons in NATO's nuclear strategy. The origins of public interest and concern over nuclear weapons in Europe lay initially in the Enhanced Radiation Warhead (ERW) episode;[22] it was intensified by the NATO decision of 12 December 1979 to introduce new long-range missiles in Europe, and then by the apparently casual references to limited nuclear war options by Reagan Administration officials. The substantial increase in public interest in, and opposition to, NATO nuclear strategy produced a somewhat ironic situation. All the time that nuclear policy was the almost exclusive preserve of government elites and a handful of interested academics, the emphasis had been on ensuring the credibility of the American nuclear guarantee: how to make American willingness to use its nuclear weapons to defend Europe credible to the Soviet Union. Yet it was the very measures that were taken to improve this credibility, such as the NATO decision with its emphasis on land-based and thus 'visible' missiles, that made public

The Reagan Administration

opinion in Europe nervous. Rather than worrying that America would not use its nuclear weapons on their behalf, a substantial number of Europeans began to worry that it would, and in a way that would effectively limit any nuclear conflict to the European theatre. Given the apparent volatility of the nuclear protector, it was not unreasonable to question the wisdom of 'coupling' Europe to the US guarantee.

Opposition to nuclear weapons need not be anti-NATO or even anti-defence *per se*, but from time to time it has assumed these characteristics when both entities are seen to be responsible for the adoption of unacceptable policies. It also has the effect of making the implementation of NATO strategy extremely difficult. As the position of the Dutch and Belgian governments regarding the NATO dual track decision has shown, the fragility of a number of governments will make their participation in any NATO programme involving nuclear modernization difficult.

Massive anti-nuclear demonstrations in several European cities appeared to confirm US suspicions that Europe was going soft, or neutralist, or both. However, in this respect, the Reagan Administration has demonstrated concern at developments in Europe. The Administration's decision to enter into arms control negotiations—first the INF talks in October 1981 and later, in June 1982, the START negotiations—was in part a response to the anti-nuclear movement in Europe. However, the beginning of arms control negotiations has only temporarily suspended the momentum of the anti-nuclear movement, and much will now depend on what progress is made in Geneva. In this respect, it is difficult to be optimistic. In view of the past rhetoric of Reagan officials on the utility of arms control and the deficiencies of previous negotiations, there can be no doubt of the Administration's sincere wish to obtain substantial reductions; the question is whether this approach is negotiable. Because of these negotiations, the NATO modernization decision and further NATO modernization proposals, the issue of nuclear weapons will certainly remain a key issue in relations between the United States and Europe. However, the degree to which the nuclear issue will be a significant factor in US-European relationships could depend on the growth and success of the Freeze movement in the United States. The tremendous increase in grass-roots support for the Freeze proposal has at least had the effect of showing many Americans that anti-nuclear demonstrations in Europe were the sign of genuine concern rather than of subservience to the Soviet Union. In this respect, it could substantially defuse much of the American sentiment over nuclear protest in Europe.

If the anti-nuclear demonstrations in Europe raised the temperature within the alliance, events in Poland took it considerably further. Following the imposition of martial law in Poland in December 1981, the Reagan Administration initiated a range of measures against the Soviet and Polish authorities, expecting the allies to follow suit. However, as with the Soviet invasion of Afghanistan, there was a difference between Washington and European capitals concerning the appropriate policies that should be adopted. European governments totally condemned events in Poland, but were not convinced that the Reagan proposals were the best way of dealing with the situation, or of obtaining improvements in the condition of the Polish people. For the American audience, any logic that lay beneath the European approach was buried by the overwhelming perception of greedy and selfish allies who put their economic well-being before the moral principle of opposing totalitarianism.

In tone and substance, American media comment of early 1982 closely resembled that of 1980, except that it was more virulent. The events in Poland and the allied refusal to support American initiatives unleashed all the underlying resentment about the allies. Among the familiar litany of complaints and criticisms, there was an increasing tendency on the part of editorials and columnists alike to question the alliance commitment. In a speech warning of the strains in the alliance, the US Ambassador to the Federal Republic, Arthur Burns, said, 'They [American troops] will not stay here if they are not welcome.' Columnist William Safire commented that, 'By their action in Poland, the Russians have put Western Europe on trial... perhaps the United States will have to assess European attitudes in plans for our own defence. We cannot defend a Europe that will not defend herself.' George Will noted that 'NATO is in imminent danger of becoming as ornamental as most modern monarchs.' Henry Brandon reported that an experienced State Department official told him: 'For the first time, the very notion of going ahead and doing things without their [the allies'] cooperation is being talked about.' The *Wall Street Journal* printed a page-long editorial proposing that the United States reconsider its commitment to Europe and turn its attention to the Pacific, and the *San Diego Union* summarized the views of many Americans when it declared, 'there are limits as to what this country can do to defend nations that refuse to be defended.'

To make things worse, not only did the Europeans fail to support the United States in applying meaningful sanctions against the Soviet Union, but they showed no sign of reconsidering their participation in the project to sell equipment for the pipeline from the West Siberian Urengoy field or other plans to purchase substantial quantities of Soviet natural gas.[23] The Administration attempted to obstruct this trade between West European countries and the Soviet Union until it was forced to back down in the face of united European opposition.[24]

The Administration's approach to the pipeline was consistent with its philosophy towards East–West relations. The predominant view in the Reagan Administration sees the West's relationship with the Soviet Union in conflictual terms: an enduring struggle which necessitates a constant search for unilateral advantage and which leaves little room for cooperation and compromise. Western policies should be coordinated within a single strategy which has the single objective of containing, or even undermining, the Soviet Union. Reagan officials appear to believe that because of the inherent weaknesses and deficiencies of the Soviet system the Soviet Union is now particularly vulnerable. This position of economic vulnerability provides the West with a classic opportunity to pressure the Soviet Union to adopt more moderate policies both at home and abroad. They argue that the Soviet leadership should not be allowed to escape the realities of the communist system through the receipt of cheap Western credit and the transfer of Western technology. Particularly important, the denial of economic help may force the Soviet Union to divert resources away from the military sector and thus moderate the Soviet military build-up. It is for these reasons that the Reagan Administration has consistently advocated a substantial restriction of Western commercial relations with the East and a reduction in the volume of trade.

This policy has found little support elsewhere in the West. European governments have accepted the need to tighten certain areas of Western economic relations with the Soviet Union, such as the export of high technology, but they accept neither the principle nor the details of the

Reagan economic strategy. The question of economic relations between East and West was discussed early in June 1982 at the Versailles economic summit and immediately after that at the NATO summit in Bonn. The NATO communiqué appeared to reconcile the somewhat different views in stating that 'economic relations should be conducted on the basis of a balanced advantage for both sides. We undertake to manage financial relations with the Warsaw Pact countries on a sound economic basis, including commercial prudence also in the granting of export credits.'[25] Only a week later, President Reagan took his allies completely by surprise by announcing the extension of the sanctions against the Soviet Union. Officials in Washington have stated that the reason for this abrupt announcement so soon after Bonn, where, on the surface at least, reason and light prevailed, was the President's personal pique at various public comments by European leaders, which he interpreted as meaning that they did not intend to impose stiffer credit terms on the Soviet Union.

If the pipeline and other controversies were not enough, relations between the United States and Europe have been further poisoned by a number of economic disputes. In theory, these problems should not affect the security relationship, since they would occur irrespective of the existence of the alliance. But, in reality, they have a very direct impact on the defence debate in two important senses. First, several European governments have argued forcefully that it is inconsistent for the United States to press the Europeans for greater defence spending at a time when American domestic policy is having such a negative effect on European economies.[26] As noted below, officials in the Reagan Administration do not accept the validity of these arguments, dismissing them as examples of European special pleading. Second, arguments over economic issues tend to revive the basic American sentiment concerning the Atlantic relationship: the fact that in the economic field the Europeans are competitors, while in the security field they continue to be petitioners. While at the official level the various US–European disputes are dealt with in virtually self-contained sectors, in the public perception they become blurred, literally producing a reaction that 'these people who are taking our jobs are the same people that we are protecting'. Economic issues thus inevitably have a serious effect on the security debate, since they hinder the creation of a climate that would facilitate reconciliation and compromise.

The Reagan Administration

The US Congress

The cumulative effect of these disputes is best measured by looking at the US Congress. In terms of its role in the US policy-making process, this is a highly significant barometer of American public opinion towards the alliance. Traditionally supportive of the principle of American involvement in the North Atlantic Alliance, Congress has always looked with caution, and not a little suspicion, at the details of this involvement. In its traditional scrutiny of US defence planning, Congress returns consistently to the argument that the United States should not protect those who are able, and yet unwilling, to protect themselves. The assumption that NATO is a gift that the United States makes to the Europeans is never far from the surface of any Congressional discussion of the alliance, and this stereotype of the Atlantic relationship is so firmly embedded that little, it seems, can change it. Even the conclusion by the Department of Defense in two comprehensive surveys of alliance burden-sharing (as discussed in the next chapter) that the allies do their fair share (or roughly their fair share) appears to have had little effect, and Congressional speeches continue to reflect the basic sentiment that the allies enjoy a free ride.

For the most part, Congress's anti-NATO sentiment remains dormant, emerging only intermittently during hearings and floor debates. Normally, a general preoccupation with domestic issues, the persuasiveness of the Administration, and a natural unwillingness to rock a well-established boat combine to contain it. Periodically, however, events serve to focus attention on the Atlantic relationship and to bring Congressional criticism to the surface in concentrated force. Developments in 1980 and 1982 caused such a resurgence. The Soviet invasion of Afghanistan, anti-nuclear demonstrators in Europe, the imposition of martial law in Poland, and the gas pipeline deal cumulatively produced a mood in Congress that was highly critical of America's allies and sceptical of the continuing relevance of NATO.

Since the days of Senator Mansfield, the seriousness of Congressional disapproval of the alliance has been measured by the support for resolutions calling for reductions in the levels of American forces in Europe. During the first half of 1982, echoes of the Senator's rhetoric could be heard quite clearly around the corridors of Capitol Hill. The

Majority leader, Senator Howard Baker, decided not to float the idea of reviving the Mansfield Amendment because, he said, 'it would start a fire I could not put out'. Senators Tower and Cohen commented that if Europe resisted modernization or increased conventional force defence outlays, 'a proposal to withdraw our troops would go through Congress like a prairie fire'. Senator Ted Stevens also reflected disillusion with the Europeans when alluding to the gas pipeline issue: 'If they [Europeans] feel so secure in their relationship with the Russians, then I think it is time for us to re-examine the number of troops we have in Europe.'

The most recent indication of Congress's mood on this issue was provided by the 12:1 vote by the Sub-committee on Defense of the Senate Appropriations Committee, chaired by Senator Stevens, which would freeze the levels of American troops in Europe at their 1980 levels, thus effectively bringing about a reduction of almost 20,000 troops. Senator Stevens pointed out that projections for FY 1983 put American force levels in Europe at 350,600, up nearly 20,000, while other NATO troop levels in Central Europe actually declined by 17,000. Officials in Washington have commented that the increases in American forces consisted mainly of support personnel involved in a number of American programmes under way in Europe, yet 4,000 of the proposed US troop reduction would be achieved by disbanding a combat brigade, the 4th Brigade of the 4th Infantry Division stationed in southern Germany. In conference, the House and Senate Committees agreed to establish ceilings which effectively mean a reduction of approximately 6,000 American troops from current levels. This is a significant indicator of the critical attitude that exists on Capitol Hill towards NATO.

In a letter to his fellow senators justifying his committee's actions, Senator Stevens stressed that the committee was not attempting in any way to weaken the American commitment to NATO. Nevertheless, he paraded the familiar complaints against the European allies, stating that 'It was time to send another signal and serve notice that the US will demand more participation from its European Allies in the defence of their own homeland ... It is not that our European Allies cannot do more. Their economies are sound and many of them have overtaken and passed the US in standards of living. Yet while we commit more than 6 per cent of our gross national product to defence, our European

The Reagan Administration

Allies remain below 3 per cent.'[27]

In a special debate in the House of Representatives, the comments were almost entirely critical of the NATO allies, to the extent that the Chairman of the Armed Services Committee, Melvin Price, felt obliged to issue a strong statement in support of the allies in order to 'permit US citizens to develop a more balanced view of the situation'.[28] Arguing strongly against the notion that the United States could force its allies to do more by threatening troop reductions, Chairman Price concluded: 'To punish our Allies is to trifle with security arrangements that have served us well for more than thirty years. Let us be firm. But let us not be foolish.'[29]

The mood in the Senate was deemed sufficiently serious to warrant several Congressional studies on NATO. The Chairman of the Senate Foreign Relations Committee, Senator Charles Percy, alarmed at the number of American commentators and legislators who were questioning some of the fundamental assumptions concerning American participation in NATO, asked his staff to conduct a review of the main problems facing NATO. At the same time, Senator Biden asked the Congressional Research Service to analyse the source of differing American and European approaches to East–West relations and the implications for NATO. While both studies agreed that the alliance faced multiple strains, they reached somewhat different conclusions: the CRS study argued that the alliance was in a severe crisis 'which could lead to a reconsideration of fundamental commitments to the North Atlantic Alliance on both sides of the Atlantic',[30] whereas the Senate study noted more optimistically that there was 'greater strength and commonality of purpose in NATO today than most political commentators believe.'[31] Motivated by similar concerns, Senator Nunn, one of the few authorities on the alliance in Congress, visited Europe during 1982 on a fact-finding mission. His report concluded: 'The NATO Alliance is now in need of major repair, militarily, politically and economically. Western political leaders must begin to make these repairs soon if the NATO shield is to continue to protect Western values and interests.'[32]

Actual troop cuts are likely to be marginal for the time being. However, there are other measures that Congress can implement which, if they lack the dramatic quality of a Mansfield-type amendment,

The Reagan Administration

nevertheless can have a serious effect on Atlantic relations and could contribute to the slow erosion of alliance cohesion. Through its role in the policy-making process, Congress can negate or redirect Administration requests concerning the American commitment to NATO. For example, Congressional committees have denied funds requested for the restationing of American troops in West Germany, insisting that the Federal Republic should bear these costs. In December 1981, the Congress passed the FY 1982 Department of Defense Appropriations Act which contained a number of restrictive amendments which seriously affected transatlantic arms cooperation.[33] These amendments were passed without debate or opposition from the Administration and involved clauses restricting the use of imported specialty metals, the purchase of Administrative use vehicles manufactured outside the United States and the cancellation of the contract to McDonnell Douglas/British Aerospace for the US navy trainer. These and minor provisions affected existing alliance arrangements to the detriment of the European allies. For example, the restriction on specialty metals is particularly serious. A number of European officials have warned that it effectively rules out transatlantic defence cooperation. The FY 1983 DoD Authorization Act passed by Congress in August 1982 repealed many of the December 1981 restrictive amendments. However, the FY 1983 Appropriations Bill passed by the Senate Appropriations Committee in September promptly reinstated a number of the restrictions.

These issues will thus remain a perpetual irritant to the functioning of the alliance. How much of an irritant will depend on a number of factors, not the least of which is the state of the US economy. Congressional debate on the restrictive clauses during consideration of the FY 1983 Authorization Act clearly reflected the growing protectionist sentiment in the Congress, especially as regards the faltering US steel and automotive industries. Cooperative arms agreements are easily portrayed as 'exporting jobs', and in the prevailing climate it could be extremely harmful politically for a legislator to be seen to be supporting a measure that could help America's allies at the expense of American economic interests. In this respect, the burden-sharing issue is never far from the scene: 'Until our allies get to the point where they are dedicating as much of their gross national product to defence as we are, it behoves this country to remember to "buy

The Reagan Administration

American" because we are shouldering the defence costs of the world.'[34]

No one can predict with any certainty how seriously Congressional criticism should be taken. Many argue that, no matter how much huffing and puffing, Congress will always shy away from voting major troop reductions in Europe. Certainly, the logic behind retaining American forces in Europe — because it is in America's own interest and because withdrawals would not save any money — is compelling. However, to argue the retention of the *status quo* because it is in America's interests is to imply a degree of rationality in the Congressional debate and decision-making process that may not always be there. Given the current levels of frustration, it is conceivable that Congress could take actions which have little to do with pragmatic reasoning and more to do with a basic desire to teach Europe a lesson.

6 Burden-sharing

At this point it would be useful to analyse the concept of burden-sharing in more detail. Use of the phrase 'burden-sharing' implies the existence of a consensus concerning what the 'burden' actually comprises; this rarely exists. If the 'burden' is defined as those resources required to ensure the security of those countries belonging to the North Atlantic Alliance, then it is evident that there are differences of opinion over what constitutes the 'security' of a nation and also what policies and capabilities are required to safeguard it. Given the diverse nature of the alliance, it is hardly surprising that there are differences over what the burden is and even less surprising that there is disagreement about how it should be divided. Even if agreement could be reached, measurement of defence effort, whether in terms of input or output, gives no indication of the relationship between that effort and the benefits a country receives in return. Furthermore, to pose the issue of burden-sharing implies an approach to alliance membership which is antithetical to the spirit of the alliance as a voluntary grouping of sovereign nations. Although it was recognized from the beginning that there should be an equitable distribution of defence tasks, it was also acknowledged that 'a final decision as to what constitutes an equitable distribution formula can never be derived from the mechanical use of statistical formulae.'[1] Thus, attempts to produce comparisons of defence effort are at best limited in application, and at worst divisive in consequence.

Despite these qualifications, in the real world leaders and officials are required to justify policies and expenditures to an ever critical electorate. For this reason the question of comparative defence effort—

or burden-sharing—has to be addressed, no matter how impracticable the result or how divisive the consequences.

The problem of measurement

Attempts to assess the defence contribution of individual members involve a complex and frequently artificial process. Measurements of defence effort can be made by examining either input, that is the amount of resources a country devotes to defence, or output, that is the actual capabilities that this expenditure buys. Measurement of input, or resources devoted to defence, can be done in a number of ways involving economic criteria. These include the total amount of expenditure on defence, the percentage of gross domestic product (GDP); the per capita expenditure; and the percentage of the national budget. Measurement of output involves a number of criteria that demonstrate how defence expenditure is spent: for example, the percentages allocated to certain key areas of the capabilities purchased in terms both of manpower and of equipment. However, as a basis for comparing defence efforts, each of these criteria suffers from limitations.

Any discussion of comparative burden-sharing must rest on data from the countries concerned, each of which has its own budget and tax system. Different methods of recruiting and managing manpower make it difficult to compare personnel costs between nations. Problems are created by fluctuations in international exchange rates and differences in the quality and use of inflation indicators. As the 1982 Pentagon Report on alliance burden-sharing states, 'NATO has attempted to deal with some of these problems, e.g. by agreeing on a common definition of what constitutes defence expenditures. NATO has not, however, formally addressed such problems as differences in purchasing power parity, the effects of taxation on defence expenditures or ways to normalize manpower costs resulting from the use of volunteers or conscripts.'[2]

Any comparison that involves currency conversion must be treated with caution because of fluctuations in exchange rates. The most obvious example of the impact of these fluctuations is the fact that when the US dollar depreciates against the currency of another ally, that ally's defence budget appears to become larger when expressed in dollars.

Similar caution should be exercised when inflation indicators are involved. Despite various efforts by the NATO International Staff to produce a formula to resolve these problems, a satisfactory solution has yet to be found. The problems of fluctuating exchange rates and inflation differentials can be avoided by using measurements such as percentage of GDP and per capita expenditure. Economic criteria of this sort, however, give no indication of the baseline capacities, i.e. the total GDP or total population of the country involved. In this sense, such comparisons give no idea of the wealth of the particular society and therefore its ability to allocate more money to defence.

Similarly, measurements of single-year performances give no indication of trends, first with regard to overall economic performance—that is, the correlation between the rise and fall of a nation's GDP and the fluctuation in its defence spending—and, second, in terms of past performance in defence spending.

Finally, indicators showing the proportion of the national budget devoted to defence are influenced by the size and shape of the respective budgets: in some countries, national budgets are 'all-embracing'; in others, government expenditure is smaller. Economic indicators also fail to reflect certain distortions due to differing tax structures. Some member countries tax their armed forces fairly heavily, with the result that a substantial proportion of their expenditure on defence returns to the Treasury.

— General economic criteria give no indication of how the money is spent on a number of hidden factors, such as:
  (i) Conscription—the low pay of a conscript results in an understatement of the real cost of conscript armies.[3] If allied manpower costs for 1979 are computed at US pay rates, the value of non-US NATO total defence would increase relative to the United States by approximately 20 per cent, reaching a total approximately equal to that of the United States. However, there are also other factors, such as associated political liability and opportunity costs, which represent political and economic burdens to those countries maintaining conscript armies.
  (ii) The involvement of the population in defence, as, for example, the extensive reserve and Home Guard systems in Scandinavia,

Burden-sharing

> which may permit a country to do more in effective combat terms with a lower expenditure.
>
> (iii) How the money is used within the defence budget, whether as in Britain, where a high proportion of defence expenditure is allocated to capital equipment, or as in Germany, where it is allocated to operational readiness, or to less relevant things in terms of combat effectiveness, such as pensions.

— Static indicators measuring output fail to account for a range of intangibles such as the training and morale of the armed forces and the quality of their equipment.

— Measurements both of input and of output fail to acknowledge the hidden costs and benefits of alliance membership involved in the provision of territory and real estate for bases and facilities at little or no cost. In this respect the cost to the Federal Republic of paying for the allied forces in Berlin are a significant yet omitted item. If the German outlay on Berlin were computed, German defence expenditures would increase by 25 per cent. When all German 'defence claims' are included—i.e. cost savings from low-cost conscript personnel, Berlin expenditures, host-nation support for both US and non-US forces, resource flows to developing countries (which can be seen as a means of supporting Western security objectives) and effects of taxation on defence expenditures—then German defence expenditure as a percentage of GDP would rise to above 5.1 per cent.

— In comparing NATO defence expenditures, it is difficult in the case of certain countries to isolate those elements that are specifically NATO-related. This is particularly true for the United States; it was true for Portugal during its military involvement in Africa; and in the same way it could be argued that the main driving force behind recent increases in the defence expenditures of Greece and Turkey has been their mutual hostility rather than their fear of the Warsaw Pact.

— It is also frequently argued that looking only at expenditure in defence represents a limited view of security. In this respect assistance to less prosperous nations which contributes to their stability is an equally important component of Western security policy. This applies within the alliance with regard to assistance to Greece, Turkey and Portugal, in which the Federal Republic has played a major part. It applies also to assistance to developing nations, a field in which the European nations bilaterally and through the European Community play a prominent role.

The Department of Defense report on burden-sharing

Following widespread Congressional criticism of America's NATO allies for their inactivity in response to the Soviet invasion of Afghanistan, the Senate passed the Levin Amendment to the 1981 Department of Defense Authorization Act. This amendment requested the Secretary of Defense to compile a comprehensive description for Congress of the contribution of the NATO allies and Japan to the common defence. The first of these reports was submitted in Congress in March 1981, and the second in March 1982.

Both reports clearly state that 'There currently exists no agreed mathematical formula that enables us to combine, with appropriate weighting, all of the major elements of burden-sharing into a precise "super indicator" of fair shares.'[4] In an effort, however, to respond to the spirit of the Congressional request for a comparison of 'fair shares ... that should be borne', and 'actual defence efforts that currently exist', the Department produced comparisons of selected quantitative indicators of member countries' *ability* to contribute to the alliance defence effort. These include a so-called prosperity index which adjusts GDP defence shares in proportion to each nation's position on the per capita GDP scale; it also compares *actual contributions* in terms of both input and output, and finally it compares the actual contributions with the ability to contribute. The same comparisons are also represented as trends.

In presenting its tentative conclusions, the Department of Defense states that it has taken into account factors that are difficult or impossible to quantify (such as host-nation support) but does not indicate how it has done this. The report notes that it has given heaviest weight to what is called the defence/prosperity index share ratio, and to a lesser degree the defence/GDP ratio, since, in the Department's view, these represent the most comprehensive indicators of defence effort and ability to contribute.

Based on these measurements, the 1982 report finds that: (a) The United States appears to be doing somewhat more than its fair share of the NATO and Japan total. (b) The non-US allies as a *group* appear to be shouldering almost their fair share of the NATO and Japanese defence burden. (c) Among the non-US nations, there appear to be wide differences regarding the amount of burden shared, with some countries doing far more than seems equitable, and some doing far less. And (d) the United Kingdom, Turkey, Greece and Portugal have a valid basis for feeling that they are doing more than their economic conditions would seem to justify. How useful is the Department's approach to resolving the question of burden-sharing, and what effect has it had?

Its chief merit is that it provides an extremely thorough review of all the quantifiable indicators that can be used to assess defence effort. However, it cannot present the whole picture. This is both because of the inherent difficulties discussed earlier in making the various calculations, such as national definitions and exchange rates, and because of the many unquantifiable or 'invisible' factors involved in alliance membership. Within these limitations and taking into account *all* its measurements, the report constitutes a useful guide to defence efforts throughout the alliance, and as such its findings are fairly predictable.

The United States, with its global responsibilities, provides a higher proportion of the total NATO defence effort than do the Europeans as a group, with their more limited horizons. Relative defence efforts among the Europeans vary over time. For example, whereas in last year's report Denmark was consistently ranked last, this year it is Norway. Perhaps the best example of the inability to tell the whole story is in the comment that both Greece and Turkey are doing more than their share, when in effect the efforts of both are directed more against each other than against the common threat.

There are several significant inconsistencies in the 1982 report. One such is its emphasis on percentage GDP as one of the two most useful measurements of defence spending. This is consistent with popular practice, since most comparisons of defence expenditure, and certainly most Congressional debates, use the percentage GDP measurement — a measurement that, not surprisingly, is in the United States' favour (the US share of GDP on defence is 5.5 per cent, whereas the NATO average is 3.5 per cent). Yet the report categorically states: 'It is insufficient to cite a single measure of performance like per cent of GDP for defence: e.g. the US contributes over 5.5 per cent compared to a non-US NATO average of 3.5 per cent. We believe that a number of measures must be considered to get a balanced picture of relative national burdens.' Consistent with this comment, the report includes a number of different indices of burden-sharing which temper the larger GDP percentage spent for defence by the United States and which, though imperfect, afford a better perspective of burden-sharing than any one single indicator.

Yet, in spite of these qualifications, the Department insists on isolating the percentage GDP measurement as particularly relevant to the burden-sharing debate. The report states that, in reaching its tentative conclusions, 'Among the ratio data, we have given heaviest weight to the defence/prosperity index share ratio and, to a lesser degree, the defence/GDP ratio, since these combine, in our view, the most comprehensive indicators of defence effort (total defence spending) and the most comprehensive indicators of ability to contribute (the so-called prosperity index and GDP).' The inclusion of the percentage GDP measurement in the 1982 report contrasts with the 1981 report, which isolated only the defence/prosperity index as being of particular comparative value.

Perhaps this new emphasis explains the slight, but politically significant, difference in the conclusions between the 1981 and 1982 reports. Despite the fact that the overall statistics of the two reports scarcely vary, the conclusion of the 1982 report is that the non-US allies are shouldering only 'roughly their fair share' of the defence burden, whereas in the 1981 report the allies were shouldering 'at least their fair share'.

A further inconsistency is that whereas the report states that its overall assessment takes account of non-quantifiable factors, it nowhere indicates how or where this is done. For example, it notes that non-military economic assistance to underdeveloped countries should be regarded as a part of a nation's overall effort. It provides statistical information on official development assistance as a percentage of GDP but does not indicate whether or how this has been taken account of in assessing and measuring member nations' defence efforts.

The 1982 report is significantly distinct from that of 1981 in tone. As the 1981 report was compiled at the beginning of the Reagan Administration, it was largely a bureaucratic effort, somewhat predictable in style, tone and substance, stating that alliance burden-sharing was more equitable than popularly imagined, but that, in view of the threat, everyone had to do better. The 1982 report, by contrast, bears the distinctive imprint of the Reagan Pentagon and is both more urgent and more critical than its predecessor.

The two differences already noted—the selection of the GDP measurement and the 'roughly fair share'—give an indication of a more critical approach. There are several other instances. For example, the report notes that the common alliance effort should not be weakened by imposing political or economic strains incompatible with peacetime objectives such as national standards of living. Yet in discussing the serious economic conditions facing all NATO nations, it comments on 'a European tendency to concentrate on social and welfare programmes to the detriment of defence expenditures. While logic may imply that those who fear war will spend more for defence rather than less, European leaders are more apprehensive of the social and political unrest which persistent inflation and high unemployment can bring.' In a similar vein, the report examines the argument put forward by some Europeans that high US interest rates have contributed to lower allocations in Europe for defence. It dismisses such notions, however: 'It is our conclusion that the high interest rate argument is not relevant to defence burden-sharing comparisons except where it is used by politicians as a rationale for non-spending in the defence arena.' Finally, it adopts a particularly critical view of the gas pipeline deal, accusing the Europeans of being swayed by economic expediency rather than defence interests.

Evaluation of burden-sharing or equity in defence effort is a highly complex and frequently subjective process. Central to any assessment of a country's defence contribution must be its economic capacity or wealth, its sense of vulnerability, its perception of the threat, and a number of extraneous factors such as cultural and historical influences. With regard to these factors, the members of the alliance operate from different standpoints, and their defence efforts vary accordingly. Defence remains a national prerogative, and although NATO develops a consensus on important policy guidelines and actions, the actual implementation of these agreed measures frequently reflects the differences in perspective of the countries concerned. Because alliance initiatives are frequently promoted by the largest and militarily most capable partner, the United States, it is not surprising that smaller nations, with fewer resources, are unwilling to implement all aspects of these proposals.

Cross-country comparison of economic data presents many problems, which are compounded if qualitative factors are also considered. The nature of the alliance militates against the production of a formula that would effectively compare defence spending or define an equitable distribution of the defence burden. Efforts to do so that are based on narrow and selective economic criteria not only fail to reflect the essential complexity of the alliance but also distort the relationship implied in membership.

7 Problems and prospects

There can be little doubt that the current tensions in the alliance will persist, at least for the lifetime of the Reagan Administration. Because security is the *raison d'être* of NATO, the question of burden-sharing will remain central to the future cohesion of the alliance. Despite this centrality, the elusive nature of the burden-sharing issue means that on its own it is unlikely to produce decisive or dramatic change. But it provides a permanent backdrop of discontent which exacerbates, and in turn is exacerbated by, other factors. In other words, the degree to which Atlantic differences over the requirements and fair share of Western defence will damage alliance cohesion depends as much on the prevailing atmosphere as on the details of the issue itself. Will the climate permit the two sides to evolve the form of quiet compromise by which differences of this nature have been solved in the past, or will emotions remain charged, thus making it difficult for the traditional mechanism of alliance diplomacy to function? In the short term, at least, it is inevitable that the economic climate and the increased attention of public opinion will both seriously limit governments' freedom of manoeuvre and, as a result, make reconciliation difficult.

Since it is highly unlikely that the Reagan Administration will retreat from its analysis of Western defence needs or that European governments will provide more resources, the gap in Atlantic perspectives of the defence burden will remain. The question is whether NATO can continue to function effectively despite this gap, and whether it can present a credible and viable defence policy that will satsify public opinion on both sides of the Atlantic.

Although this section will concentrate on the military options

Problems and prospects

available to NATO, it should be noted that public support for future NATO defence policy depends on the alliance maintaining an equal emphasis on détente policies. Public support can be generated for defence spending for necessary modernization projects, but only if these are accompanied by parallel efforts in the political field to achieve greater stability through negotiations.

More of the same

The most likely course of action for NATO to adopt will be a 'more of the same' approach. Within the traditional limitations of structure and resources, the alliance will continue to explore every avenue in order to enhance the credibility of NATO's defence. This search will involve such time-old formulae as division of labour, rationalization, specialization, etc. But underneath these initiatives the charade of agreeing to disagree will continue. As in the past, the United States will continue to harass and pressure the allies to provide more resources and take on more responsibilities. The Europeans will continue to endorse American strategic assessments, but will accept the consequences of these assessments only inasmuch as they fit in with national plans and do not involve spending more money. How long this situation will endure will depend on the degree to which the American expectations are assuaged by the various initiatives under way and the extent to which defence is able to escape the glare of public scrutiny.

In order to satisfy American criticism, great emphasis has already been placed on the concept of the 'division of labour' as a means of strengthening alliance security. But this concept raises a number of important questions for alliance planners—principally that neither the objectives of the division of labour concept, nor the responsibilities or resources that it should involve, have been satisfactorily defined. Is the objective to give visible proof of alliance cohesion in sharing the defence burden equitably, or is it to make the most rational use of existing resources?.While many observers would answer both, in fact the two are not necessarily compatible. It may make sense for symbolic reasons to encourage the Europeans to make military contributions to the Gulf, but is it the best use of resources when European commitments can have only marginal military significance? How important is

Problems and prospects

it that Europeans are seen to be 'alongside' the United States in its Gulf commitment? Surely it is preferable that Europeans should use their resources in the most efficient and cost-effective way, despite the fact that the resulting measures may lack the visibility necessary to satisfy American criticism.

In view of the political and economic pressures threatening NATO strategy and cohesion, there were expectations that the Bonn summit in June 1982 would yield a new initiative. Since this was President Reagan's first NATO summit, many observers forecast a new US initiative, possibly on the lines of the Carter LTDP. In the event, the initiatives were very much a continuation of past policies and practices. However, if the recommendation for action had a familiar ring, nevertheless there were several specific references that indicated a significant trend in alliance thinking. The Final Communiqué, signed by all Heads of State, contained the commitment to 'seek to achieve greater effectiveness in the application of national resources to defence, giving due attention to the possibilities for developing areas of practical cooperation. In this respect, the Allies concerned will urgently explore ways to take full advantage, both technically and economically, of emerging technologies.'[1]

The accompanying 'Document on Integrated NATO Defence' laid particular emphasis on improving NATO's conventional forces: 'We will continue to strengthen NATO's defence posture, with special regard to conventional forces.'[2] It outlined several areas that would receive renewed attention in the fulfilment of NATO force goals, including measures to improve the readiness of the standing forces and the readiness and mobilization capability of reserve forces; implementation of LTDP measures; a continued search for ways to achieve greater effectiveness in the application of national resources with due attention to fair burden-sharing and to possibilities for developing areas of practical cooperation; and exploitation of new technologies. Promises of fulfilment of force goals, of better cooperation and of a more effective use of resources are a well-established part of any NATO communiqué. However, the specific reference to the need for improved conventional forces, for the urgent exploitation of new technologies and for improved reserve forces, although not entirely new, certainly indicate a change of emphasis.

The conventional option

Interest in improving NATO's conventional forces is hardly original, for alliance officials have long emphasized the importance of the conventional arm of the NATO triad (comprising strategic nuclear, theatre nuclear and conventional forces) in raising the so-called nuclear threshold. However, the coincidence of two trends has caused the spotlight to be cast on the role of conventional forces. First, the widespread public unease over the role of nuclear weapons in NATO strategy which not only threatens the deployment of the new intermediate range systems, but also questions the practical utility of NATO's battlefield nuclear systems; and, second, the emergence of new technologies which suggest that missions which previously involved a reliance on nuclear systems can now be carried out with conventional means. Under these pressures, it was inevitable that NATO would have to give some thought to improving its conventional forces.

The case for reducing NATO's reliance on nuclear weapons and the need to increase conventional capability was given prominence in an article by four influential former US officials. Writing in *Foreign Affairs* in spring 1982, Messrs Bundy, Kennan, McNamara and Smith proposed that the United States renounce 'first use' of nuclear wepons.[3] They argued that first use lacks military or political advantage and that a more robust conventional deterrent would be both adequate and affordable. A response by four German authors, while disputing the advisability of a declaratory policy of 'no first use', argued in favour of raising the nuclear threshold by strengthening conventional defence capability.[4]

The debate over a conventional defence for NATO has been joined from several quarters, official and unofficial. Senator Nunn, noting NATO's dilemma of having an incredible nuclear strategy and inadequate conventional forces, called for conventional force modernization as the only solution.[5] Similarly, Manfred Wörner, before becoming West German Defence Minister, released a study which argued along similar lines.[6] Both men proposed that the alliance exploit modern weapon technology to raise the nuclear threshold and allow for a substantial reduction in the number of short-range nuclear weapons deployed on West German territory. General Rogers appeared to give this notion official blessing when in September 1982 he stated that it would be possible to withdraw many of the 6,000 nuclear warheads now based in Western Europe if sufficient conventional forces could be fielded. More significant, he put such a conventional capability within the realm of the achievable by stating that it would require a 4 per cent increase in real defence spending per year over the remainder of the decade.

The question of improving NATO's conventional forces has become a highly topical issue and there has been a proliferation of study groups and articles seeking to establish whether a credible conventional defence is feasible and affordable. There is general agreement that NATO has to do something; the question is what, and at what cost. What are the priorities: should efforts be directed at achieving overall improvements or should they aim at specific areas of greatest weakness?

Assessments of what NATO needs in order to provide a credible defence depend on estimates of the current balance of forces in Europe and the level of insurance required. Both criteria derive from assumptions that are, for the most part, subjective. There is no way of reliably knowing when nations would mobilize or for how long, what forces would be available or how they would fight. Comparisons of individual weapons systems have little utility when they are used in different force structures for different missions and by soldiers of varying experience, and when they do not normally confront each other in a one-to-one situation. The forces that NATO nations require can only be the product of a balanced assessment that is based on a range of factors, military, political and economic. The military provide their requirements based on a broad range of contingencies, but it is for political leaders to take these military requirements and assess them within the prevailing political and economic environment. Two very different articles offer examples of the difficulty of establishing NATO's real needs.

In an article written in summer 1982, *The Economist* provided an assessment of what NATO would require in the way of conventional forces in order to reduce its reliance on nuclear weapons. Assuming that in a non-nuclear conflict the Warsaw Pact would currently break through NATO defences within a matter of days, or at least weeks, it asked: 'How many extra men, tanks, guns and aircraft would NATO need to be able—probably—to stop them winning, and therefore—probably—to deter them from even trying.'[7] In its analysis, *The Economist* notes that although the force ratios in Central Europe are not so disadvantageous for NATO as commonly supposed, the Warsaw Pact enjoys several important advantages, such as the high proportion of Soviet troops, better combat to support ratios, and geographic proximity giving easier mobilization. In assessing NATO's requirements, the article assumes that NATO nations will fulfil the commitments made within the 3 per cent pledge, and then estimates what extra units NATO would need to fight 'a solid defensive battle on the central front for 30 days'. It reaches the conclusion that these additions could be achieved by an additional real increase of approximately 1-1½ per cent on top of the 3 per cent. If the extra measures required were put into effect next year, *The Economist* calculates that this would mean the Federal Republic spending around 4.2 per cent of its GNP on defence by 1986 (compared with 3.4 per cent now), France 5 per cent (compared with 4.2 per cent) and the United States 7.5 per cent (compared with 5.8 per cent). It points out that these calculations are based on a 30-day war, including only the immediately available standing forces of either side and their early reinforcements. A conflict that extended beyond that period would be considerably to the Soviet Union's advantage.

As an analysis, the *Economist* survey suffers from a number of serious limitations. It does not explain the basis for its calculations either of numbers or of cost, nor does it provide any rationale for the numbers it believes necessary for credible deterrence. It is in effect a broad-brush approach to the problem, one that relies on creating the appearance of credible deterrence, through the provision of greater numbers and the creation of a balance of forces that does not provide the attacker with any obvious advantage, or at least contains sufficient uncertainties and imponderables to make him think twice.

By way of contrast, an American analyst, Edward Luttwak, in a severe attack on the 'no first use' recommendation, pours scorn on the notion that under present circumstances NATO could mount a credible conventional defence.[8] He notes that in terms of raw numbers, manpower, GNP, etc., the picture for NATO is reasonably attractive. However, if the focus is narrowed to the Central Front and to comparisons of individual weapon systems and operational details, then the situation for NATO rapidly becomes grim. Luttwak comments that, with the resources it has at its disposal, NATO could have larger, better equipped forces, but that 'those in Europe who understand such matters know that an increased effort would not improve the balance unless it were truly huge because there are two fundamental factors at work ... which would nullify the benefits of any marginal increase in defense spending, just as they outweigh every one of the disadvantages that afflict the Soviet Union.'

The first of these factors is the fact that NATO is a defensive alliance, not just in declared intent, but in actual military orientation. This means that the Soviet high command can concentrate its forces at will. Luttwak calculates that, excluding East European divisions and allowing for Soviet forces to cover regional contingencies (China, the Southern Front, etc.), this advantage would allow the Soviet Union, upon mobilization, to launch 80 divisions against NATO in Central Europe. He notes that the International Institute for Strategic Studies gives total mobilization figures of 118 for the Soviet Union and 116 for NATO, both of which figures he dismisses as too high. The Soviet figure he reduces to 80 by making more conservative allowances for regional threats. But he reduces the NATO figure to 35 divisions—first, by discounting the US National Guard division as unavailable; second, by taking out those NATO forces belonging to Southern Flank countries; and finally by comparing the 58 remaining NATO divisions with Soviet division equivalents. Thus, he assesses that in reality 35 NATO divisions would face 80 Soviet divisions. However, the true combat imbalance would be ever greater, since the Soviet divisions can be concentrated during an offensive against a few narrow segments of the front, whereas NATO's divisions must defend all along the 400-km border.

The second decisive factor in Luttwak's analysis is the nature of armoured warfare and the fact that the Soviet army is, in his words, 'the only one which stages vast army-sized exercises to educate its officers and men in the broad art and detailed craft of armoured warfare'. He paints a vivid picture of a Soviet armoured attack, which he postulates would not be a set-piece offensive on a pre-planned line of advance, but rather would seek to advance opportunistically. Advance regiments would probe for gaps, and reinforcements would be sent to add to the momentum. The Soviet army would achieve a blitzkrieg effect without the skill and initiative of regimental officers. 'By feeding reinforcement echelons into avenues of penetration successfully opened

Problems and prospects

... theirs would be an advance ... achieved by mass and momentum from above.' In short, Luttwak concludes that the crucial difference is that the Soviet army has a valid method of offensive war, whereas NATO, for its part, has no valid method of defence. In the face of these conditions, he argues that NATO could mount a successful conventional defence only if it were willing to attack first in order to disrupt an offensive pre-emptively, or if the defender had geographical space in which to manoeuvre and fight a defence-in-depth strategy[9] — both ideas he notes are politically untenable for NATO.

Luttwak's analysis provides an interesting contrast to that of *The Economist* because he insists on looking beneath the numbers and on examining what the numbers mean in terms of operational detail. For example, he refutes the claim that NATO could effectively utilize new precision-guided anti-tank weapons to counter the Soviet armoured threat. He argues that if one looks at the total combat picture, it is evident that tank missile crews would find it difficult in combat conditions to maintain the necessary line of sight, and that they would be neutralized by the overwhelming power of Soviet artillery. Nevertheless, although his analysis provides a refreshing and sobering corrective to the somewhat casual and unquestioning acceptance of force ratios based solely on broad numbers, as a reliable guide to force requirements it is equally flawed. Most significantly, it follows a worst-case pattern, in which it is assumed that everything will go right for the Soviet Union and wrong for NATO. His assessment of initial force ratios is based on a scenario of total mobilization which gives the Soviet Union maximum advantage. (This is consistent with his general theme that since the Soviet Union is aggressive and NATO defensive, the Soviet Union can choose the scenario that suits it most even if this means a lengthy mobilization period.)

Luttwak's description of a smoothly-oiled Soviet offensive machine grinding relentlessly through the disorganized NATO forces appears, to say the least, unduly pessimistic. The characterization of Soviet armed forces is hardly consistent with what is known of their invasion of Czechoslovakia, where, despite being unopposed, they did not perform like clockwork, or the invasion of Afghanistan, where reports stress the slowness of the Soviet military system to adapt to local conditions. Neither of these are necessarily analogous to an offensive against

Problems and prospects

Western Europe, but they do give some indication of the limitations of Soviet forces. Furthermore, criticisms contained in Soviet military literature concerning the performance of Soviet soldiers suggest that a Soviet armoured thrust may not be quite so overwhelming as Luttwak suggests. Given the unwieldy nature of modern armies and the attendant command and control problems, one wonders if a Soviet commander would view his chances of success quite as optimistically as Luttwak does.

The most useful aspect of Luttwak's analysis is the light it throws on the somewhat blurred relationship between military and political assumptions in force assessment, the shadowy area of intentions and capabilities. His study is an analysis of military factors, of how Soviet and NATO force ratios would interact in operational terms, yet beneath these assessments lie a number of political assumptions concerning Soviet intentions. In Luttwak's view, the Soviet Union is an aggressive power seeking to expand its influence whenever possible; it has developed offensively orientated forces to support these objectives and will use these forces under conditions of maximum advantage wherever it can. NATO must therefore prepare for the most likely and dangerous threat, a fully mobilized Soviet offensive. Yet many observers find the concept of a conventional war in Europe, following a lengthy build-up, difficult to accept because it is not consistent with political and economic realities. It is a scenario that occupies the extreme end of the threat spectrum and should be judged accordingly in NATO's planning priorities. Since it is difficult to identify with Luttwak's political assumption, it is difficult to accept the consequent implications of his analysis.

These two studies demonstrate alternative approaches to solving NATO's problems: the first involves general improvements to all areas of NATO's forces, concentrating on improving existing force ratios and capabilities, but involving no radical change in existing NATO force posture or doctrine; the second suggests that a conventional defence can be obtained only through radical changes in doctrine and capability. NATO must exploit new technologies to obtain the conventional means to strike and disrupt Warsaw Pact forces early in any conflict and must reorganize its ground forces to provide defence in depth.

The most practical discussion of NATO's current malaise, and one which appears to offer a middle way between the two alternatives, is

Problems and prospects

contained in the Nunn Report. Senator Nunn argues for substantial change but believes that such changes are affordable. In his report he describes several new initiatives which could make a credible conventional defence possible. These include the exploitation of new technologies, the adoption of new operational doctrines and better utilization of Europe's substantial reserves of trained manpower.

Central to Senator Nunn's proposals is the Airland Battle concept, as described in a pamphlet published by the US Army.[10] Airland Battle outlines the possibility of successfully countering a Warsaw Pact conventional attack without having to resort to the early use of nuclear weapons. The concept foresees early attack with conventional munitions on Warsaw Pact reserves at the rear. Successful disruption of these Warsaw Pact reserve forces should in turn make the task of engaging front-line units more manageable. This emphasis on early attacks against enemy rear forces would be accompanied by a greater emphasis on defence in depth in order to cope more effectively with Warsaw Pact attacking forces. As Senator Nunn points out, Europe has large reservoirs of trained manpower that—if organized and re-equipped into standing reserve units—would provide defence in depth.

Airland Battle is only one of several concepts currently in circulation that would involve the utilization by NATO of conventionally armed missiles, ballistic and cruise, to perform operations previously carried out either by aircraft or by nuclear weapons. These concepts capitalize on four major advances in conventional weapons technology: improved guidance systems; greater ability to collect, process and distribute information; improved mobility and target surveillance; and tremendous improvements in small munitions. Proponents argue that full exploitation of this technology by NATO would allow the alliance to delay, disrupt and hold a Warsaw Pact conventional offensive without recourse to nuclear weapons.

While in theory these ideas offer considerable potential to NATO, little thought has been given about how they would be applied on an alliance-wide basis. There are many problems. The new concepts are, for the most part, US-inspired, and have not yet been debated within NATO. Adoption by other members would require agreement that this was the most appropriate way for the alliance to spend its resources; that the system could be afforded; and that new procurements could be

Problems and prospects

accommodated within national planning cycles. Inevitably, some of the assumptions on which the initiatives are based will be questioned, as will the idea of placing such reliance on relatively new and untried technology. These concepts are based on technology currently evolving in the United States. While European companies may have similar developments under way, the adoption of these ideas by NATO would suggest large-scale purchases by the Europeans from the United States.

There could also be political objections. Deployment of land-based missiles designed to strike early in a potential conflict against Warsaw Pact territory could be seen to have 'offensive' connotations and to represent a turning away from NATO's traditional posture as a defensive and reactive alliance — a change in direction that would not be supported by political opinion. Indeed, in the current political climate, it is difficult to envisage public support for the deployment of new missiles even if they were conventionally armed. It is also unlikely that the emphasis on defence in depth which is explicit in Airland Battle would be acceptable to any German government.

The great obstacle to the adoption and implementation by NATO of these initiatives is the propensity of the alliance to stick to the *status quo*. In Senator Nunn's view, conservatism rather than cost is the real problem: 'Many of the new efforts do not require additional money, but instead focus on reorganization and revitalized strategic and tactical doctrine. The primary cost of such change would be in shattered preconceptions and broken traditions.' However, having recognized the nature of the problem, Senator Nunn is concerned about the urgency of the situation and argues that the United States must now exert serious pressure on its allies. He consequently recommends:

> Until the Alliance agrees on a new doctrine and dedicates itself to its implementation, President Reagan should instruct the Secretary of Defense to permit no net increase in the present number of US troops deployed in Europe . . . and beginning in 1984 to isolate those expenditures [in the five-year defence plan] designed directly for NATO improvements and freeze those expenditures which exceed the NATO 3 per cent commitment until the NATO political leaders have a clear agreement on a military strategy for the decade of the eighties.[11]

Problems and prospects

The fact that a senator regarded as a supporter of NATO would recommend pressure of this nature is an indication of the seriousness with which the current situation is viewed on Capitol Hill.

Current pressures clearly call for initiatives to improve NATO's conventional forces, and NATO ministers have indicated that such initiatives are under review. However, the question remains, what should be done and at what cost? As previous discussion has demonstrated, there is no magic formula to determine what NATO needs to provide a credible conventional defence; it is a question of judgment. Many argue that deterrence can be assured simply by improving the general appearance of defence capabilities, by utilizing emerging technology where appropriate and affordable in order to ensure steady rather than radical modernization of conventional capabilities. According to this argument, the current political environment requires continuity rather than radical and potentially destabilizing change. Furthermore, to ensure public support, modernization must be accompanied by efforts to control and limit armaments and produce greater stability through negotiation. Others, however, would argue that NATO's problems are now so serious, its military deficiencies so great, that only radical changes in capability and doctrine will maintain its future credibility. A major alliance-wide initiative is required that will effect a quantum jump in capability.

Two criteria will determine whether the alliance adopts a 'more of the same' or a quantum jump approach: affordability and perception of the threat. It is a measure of the gap between the United States and Europe that affordability will be the prime determinant for the Europeans, whereas the Soviet threat will be the main consideration for the United States.

The European option

The alliance faces multiple problems. The central dilemma confronting NATO governments is to maintain the credibility of NATO's deterrent posture in a manner that is compatible with available resources and which commands public support on both sides of the Atlantic. NATO strategy must be credible and viable. Yet it must satisfy these criteria during a period when no additional funds can be expected, when

Problems and prospects

existing expenditure buys less and less capability, when additional tasks are being proposed, and when the modernization of Soviet armed forces continues unabated. Most significant, it must do so against the background of an eroding public consensus for alliance defence policies. The problem is that NATO strategy must respond to three different audiences: it must deter the Soviet Union, protect and reassure European public opinion, and satisfy the frustrations of the US Congress. The requirements of these audiences are not easily reconciled.

The future credibility and cohesion of the alliance will depend on the resolution of three issues: the role of nuclear weapons, the related question of more effective conventional forces, and the issue of 'out of area' responsibilities. Each of these problems reflects the problem of implementing NATO's strategy of deterrence through flexible response. A strategy based on maintaining flexibility of action represents the only acceptable policy for an alliance comprising such divergent interests. Yet it poses a number of problems. The necessity to close off every potential gap or weakness makes the selection of priorities difficult, and provides a permanent source of disagreement about 'how much is enough' at any particular level. In the absence of extra resources or a 'quick fix' to make better use of existing resources, a choice will inevitably have to be made between, on the one hand, selecting priorities, such as readiness over sustainability, in which case the strategy becomes less flexible, and, on the other, continuing to provide the current range of capabilities but accepting a lower margin of insurance at each level.

The NATO INF decision will continue to be the dominant issue in alliance politics, not only because of the political capital invested in the double track decision but because of the fundamental questions it raises about the requirements of extended deterrence and the role of arms control in alliance security policy. In addition, the US Congress will be looking for tangible evidence that the Europeans are prepared to make additional efforts for the defence of Western interests. Hence there will be considerable interest to see what support the United States receives in its plans for the Gulf. Similarly, Congress will want to see whether the Europeans are willing to solve the current dilemma of NATO strategy by finding the extra resources necessary to take full advantage of emerging technology. Unfortunately, symbolic gestures and grand initiatives do not always make for the most effective use of resources. Likewise, pragmatic defence planning through the NATO mechanism lacks the visibility necessary to provide proof of extra commitment.

In consequence, a gradual erosion of alliance cohesion is threatened. The current differences between the United States and Europe are so profound that the traditional alliance technique for reconciliation may not be sufficient to contain them. However, rather than stumbling from crisis to crisis, another route has become evident. As the transatlantic gulf has widened over the past two years, murmurings have been heard from many quarters in Europe for the need for Europeans to develop and assert a common perspective on the requirements of Western security. A collective and independent European voice would thus be established to counter the weight of the United States. Support for this concept has been found right across the political spectrum, although the motivation and the degree of independence sought varies considerably: many of those in the anti-nuclear movement would like to see an independent Europe distance itself from the dangerous and irrational policies of the United States and establish a 'third way'; others see a more assertive Europe as a means of strengthening the alliance by providing the long-awaited 'second pillar'; while some conservatives see the development of an independent European defence entity as a necessary preparation for the inevitable US detachment from Europe and as a counter to the forces of neutralism and pacificism. Thus, irrespective of motivation, there is considerable support for the development of greater consultation, cooperation and harmonization among European nations on the question of Western security policy.

As a means of countering current trends within the alliance, a unified and coherent European voice could be helpful in a number of ways. It could help to clarify a number of existing differences. Rather than endorsing American initiatives which they have little hope (or even intention) of fulfilling, it would be preferable if the Europeans would explain and justify why they do not accept American assessments and the requirements that flow from them. A collective European voice could influence the United States, as regards both military and foreign policy issues, to pursue policies that are more in line with political views in Europe. It could also convince an ever sceptical Congress that Europeans are prepared to think responsibly about their own security —

Problems and prospects

even if the answers are not always to Congress's liking. In Europe a more visible and assertive European influence in alliance security policy could help to sustain public support for defence policies. Implicit in many of the anti-nuclear demonstrations is a resentment that alliance policies are imposed by the United States — witness many of the criticisms of the NATO decision to deploy new intermediate range nuclear missiles in Europe. Although, ironically, in this particular issue the criticisms are not justified, it is nevertheless inevitable that, because of its size and the resources it can muster, the United States plays a dominant role in alliance planning. Finally, of course, as commentators never tire of pointing out, it is time that Europe developed military and political responsibility commensurate with its economic status.

However, while in theory the evolution of a more independent and unified European voice makes eminent sense, in practice the formidable obstacles that have always precluded such a development remain. First, there is the institutional question: which institution should provide the basis for such a development? The three existing institutions — the Western European Union (WEU), the European Community (EC) and the Independent European Program Group (IEPG) — all have serious limitations as regards membership and influence.

As the original institution formally credited with responsibility for the defence of Europe through the Brussels Treaty, the WEU would appear to have strong claims as the focus for the development of a common European approach to security. This case appeared to be strengthened when, during 1981, the French government began a campaign to revitalize the WEU, and several French ministers and deputies made speeches in which they specifically referred to the need to strengthen it. Despite this burst of enthusiasm, however, there has been little follow-up. Some observers believe that the French government was using the opportunity to emphasize and reinforce their commitment to Western Europe's defence rather than actually promoting the WEU itself. Despite its strong legal claims, the cause of the WEU finds little support outside France. The common view is that its membership is too restrictive and that it is in many senses a redundant organization. To coordinate European security policy through the WEU would be to duplicate NATO while relinquishing the essential political basis of the European Community.

The IEPG also has potential as a means to achieve greater coordination among Europeans on security matters. The IEPG was established in 1976 as the forum in which European nations, including France, could coordinate equipment procurement. It was intended to represent the European end of the two-way street that many hoped would materialize with the United States. Its success in producing collective projects has been extremely limited. However, it does provide a useful opportunity for European defence officials to meet on a regular basis, and there would appear to be no reason why its mandate could not be expanded to include broader questions than equipment procurement. However, participation would have to be considerably upgraded, and its organization given a more permanent character. The IEPG could offer a 'quick fix' to the problem of harmonizing European views, but it would not provide a long-term solution, since it lacks any political structure.

In the final analysis, the only institution which offers long-term potential for presenting a coordinated European framework is the European Community. The idea that the European Community should embrace the concept of European security has long been advanced. However, there have always been and still are powerful influences that oppose such a development.

First, there is the current condition of the EC itself. Many believe that the Community has more than enough problems on its current economic and social agenda, without further burdening it with sensitive questions of security. Furthermore, with the accession of Spain and Portugal looming in the future, its problems will certainly increase, and its cohesion will be placed under further strain. Yet some observers argue that it is precisely because movement to closer integration on the economic front appears to be stalled that it is important to pursue cooperation in the security field. Such a development could provide the momentum towards integration that is currently lacking, particularly by reminding the European public of its common perspectives and common interests when considering the world at large.

The second major inhibition to Community involvement with security policy concerns membership. The fact that neutral Ireland is a member of the EC and that several European members of NATO are not members of the Community (currently, Spain, Portugal, Norway,

Problems and prospects

Turkey and Iceland) has always been regarded as a serious obstacle to the effective development of a European security policy.

Yet, despite these inhibitions, considerable progress has already been made that could provide the basis for the Community to move forward into the area of security cooperation. Through European Political Cooperation and the European Council, members of the Community have made very substantial progress in the coordination of foreign policy.[12] Harmonization of foreign policy exists over a wide range of issues, and Community ambassadors and officials meet regularly in a variety of forums. It should be noted that, at this stage, cooperation involves the harmonization of policies rather than their initiation. Nevertheless, the Community has managed to present a distinctive European approach on issues such as the Middle East, Afghanistan and Poland. As Christopher Tugendhat has recently commented: 'What could be more natural now than for the ministers meeting in Political Cooperation and the European Council to devote an increasing amount of time and attention to such matters as the proper balance between military capability and arms control, the need for the most cost-effective use of defence resources and questions concerning the deployment of particular types of weapons.'[13]

Stimulus for the Community to move in this direction has been provided by the foreign ministers of the Federal Republic and Italy in what is known as the Genscher–Colombo Plan. While this plan to revitalize the concept of European union has three parts, covering economic, cultural and political issues, it is the last that has achieved most attention because of its relevance to security cooperation. In fact, it was security cooperation which provided much of the motivation for the plan, and Foreign Minister Genscher devoted the bulk of his speech before the European Parliament to security cooperation. The plan calls for 'the coordination of security policy and the adoption of common European positions in this sphere in order to safeguard European independence, protect its vital interests and strengthen its security.' It met with substantial criticism when it was presented to the European Parliament, with Danish and Irish members and French Communists and Gaullists all expressing strong reservations. At the governmental level, the Irish and Danish governments consistently oppose moves to extend the Community's mandate to the field of security.

Problems and prospects

Nevertheless, the plan has gained some important support. The former Belgian Premier, Leo Tindemans, wrote a report five years ago which urged a similar initiative and commented: 'If Europe wants to have common external policies, it must have the courage to have ideas on its own security.' The European Parliament is also showing signs of interest in this area. Security has already been the subject of parliamentary reports over the Gladwyn and Klepsch reports, but now the directly elected parliament will shortly receive a report on European security questions by a Danish Liberal, Niels Haagerup, and a further report on European cooperation in equipment procurement from Mr Adam Fergusson. The President of the European Parliament, Mr Pieter Dankert, a former spokesman on defence and foreign affairs for the Dutch Labour Party, has also indicated his belief that the issue of European Security should be given greater visibility.[14]

Despite these signs of progress, many problems exist to inhibit the development of a common European approach to security. Noting the pressures that are causing many Europeans to think in these terms, an American analyst compared the European predicament to that of a trapeze artist, wanting to jump but afraid to let go. However, it could be that events will remove the element of choice. Current American policy may turn out to be more than a temporary bout of 'Reaganitis' and could represent a more permanent shift in American attitudes and policy. If this is the case, then the familiar refrain that common standards and objections will always overcome Atlantic differences will no longer be valid and the Atlantic rift will surely become permanent.

Under these circumstances, the requirement for a coordinated European approach will become even more urgent, and the questions that are now being asked under the shelter of American military protection will have to be asked in the shadow of Soviet military power. Europeans will have to consider many of the issues that the existence of NATO has long freed them from — questions such as the degree of military insurance required, the role of nuclear weapons, and the nature of security threats beyond the continent's boundaries.

Tables

Table A Selected indicators of ability to contribute, 1980

	Share of total GDP		Share of total population		Per capita GDP as % of highest nation		Adjusted GDP share*	
Belgium	1.76%	9	1.42%	11	86.0%	5	1.98%	9
Canada	3.78%	7	3.44%	8	75.9%	9	3.76%	6
Denmark	0.98%	10	0.74%	13	92.2%	3	1.18%	10
France	9.64%	4	7.73%	6	86.4%	4	10.92%	4
Germany	12.11%	3	8.86%	3	94.7%	2	15.04%	2
Greece	0.60%	13	1.37%	12	30.2%	13	0.24%	12
Italy	5.82%	6	8.21%	4	49.2%	12	3.75%	7
Luxembourg	0.07%	15	0.05%	15	85.8%	6	0.08%	15
Netherlands	2.48%	8	2.03%	9	84.4%	7	2.74%	8
Norway	0.85%	11	0.59%	14	100.0%	1	1.11%	11
Portugal	0.36%	14	1.43%	10	17.2%	14	0.08%	14
Turkey	0.84%	12	6.49%	7	9.0%	15	0.10%	13
UK	7.74%	5	8.06%	5	66.5%	10	6.75%	5
US	38.19%	1	32.75%	1	80.8%	8	40.46%	1
Japan	14.80%	2	16.83%	2	60.9%	11	11.81%	3
Non-US NATO	47.01%		50.42%		64.6%		47.73%	
Non-US NATO + Japan	61.81%		67.25%		63.7%		59.54%	
Total NATO	85.20%		83.17%		71.0%		88.19%	
Total NATO + Japan	100.00%		100.00%		69.3%		100.00%	

*These statistics are obtained by multiplying each country's share of total GDP (column 1) by its per capita GDP expressed as a percentage of the highest per capita GDP (column 3), and then expressing each result as a percentage of the total. The purpose is to present an indicator of GDP share adjusted for the differing levels of prosperity among member countries. *Source: Report on Allied Contributions to the Common Defense*, op. cit. (ch. 6, n. 2), pp. 21–3.

Note: The difficulties involved in making cross-country comparisons are discussed on pages 54–5. A more rigorous method than that used in the *Report* produces significantly different figures for the 'adjusted GDP share', though it affects the ranking only slightly. For example, the alternative method results in adjusted GDP shares of 8.41% for France, 10.5% for Germany, 4.98% for the UK, 46.9% for the US and 14.28% for Japan.

Table B Selected indicators of contribution, 1980

	Share of total defence spending		Defence spending (% change 1971–80)		Share of total active defence manpower		Active defence manpower (% change 1971–80)		Share of total active and reserve defence manpower		Share of total ground forces ADEs		Share of total tactical combat aircraft	
Belgium	1.52%	9	49.7%	5	1.57%	11	1.3%	7	2.00%	11	1.81%	12	2.96%	9
Canada	1.90%	8	6.0%	11	1.61%	10	-7.2%	10	1.15%	13	0.85%	13	2.53%	10
Denmark	0.62%	13	3.6%	12	0.59%	14	-18.5%	12	1.11%	14	2.25%	10	1.33%	12
France	10.13%	4	36.1%	6	9.67%	3	0.9%	8	9.58%	4	4.82%	8	8.11%	4
Germany	10.23%	3	23.3%	7	9.06%	4	3.3%	5	11.67%	3	10.74%	3	8.85%	2
Greece	0.87%	11	77.9%	3	2.86%	8	3.9%	4	4.29%	7	5.13%	7	4.01%	7
Italy	3.67%	6	15.4%	9	7.50%	6	-8.0%	11	7.77%	5	6.48%	4	5.05%	6
Luxembourg	0.02%	15	73.3%	4	0.02%	15	8.3%	2	0.01%	15	(a)	15	0.00%	15
Netherlands	2.02%	7	10.3%	10	1.82%	9	-5.8%	9	2.45%	10	3.03%	9	2.43%	11
Norway	0.64%	12	20.6%	8	0.68%	13	7.2%	3	2.47%	9	2.22%	11	1.23%	13
Portugal	0.33%	14	-23.6%	15	1.35%	12	-60.1%	15	1.22%	12	0.49%	14	0.69%	14
Turkey	1.02%	10	105.6%	1	10.51%	2	18.9%	1	12.36%	2	11.99%	2	3.91%	8
UK	10.29%	2	3.3%	13	7.86%	5	-19.5%	13	6.28%	6	5.32%	6	8.61%	3
US	52.97%	1	-11.4%	14	41.30%	1	-20.7%	14	35.13%	1	38.62%	1	44.72%	1
Japan	3.77%	5	78.8%	2	3.60%	7	2.3%	6	2.51%	8	6.24%	5	5.57%	5
Non-US NATO	43.26%		19.9%		55.11%		-4.7%		62.36%		55.13%		49.70%	
Non-US NATO + Japan	47.03%		23.2%		58.70%		-4.3%		64.87%		61.38%		55.28%	
Total NATO	96.23%		0.4%		96.40%		-12.3%		97.49%		93.76%		94.43%	
Total NATO + Japan	100.00%		2.1%		100.00%		-11.8%		100.00%		100.00%		100.00%	

(a) = less than 0.005%. ADE = Armoured Division Equivalent

Table C Selected indicators comparing contribution with ability to contribute, 1980

	Ratio: Def. spend. share/GDP share		Ratio: Def. spend. share/Adjusted GDP share*		Ratio: Active def. manpower/Pop. share		Ratio: Active and res. def. manpower/Pop. share		Ratio: ADE share/Adjusted GDP share*		Ratio: Aircraft share/Adjusted GDP share*	
Belgium	0.86	7	0.77	8	1.11	6	1.41	5	0.91	9	1.49	4
Canada	0.50	13	0.51	13	0.47	13	0.33	13	0.23	14	0.67	12
Denmark	0.63	11	0.53	12	0.80	12	1.50	4	1.91	5	1.13	7
France	1.05	5	0.93	7	1.25	4	1.24	7	0.44	13	0.74	11
Germany	0.84	8	0.68	10	1.02	7	1.32	6	0.71	11	0.59	13
Greece	1.45	1	3.63	3	2.09	1	3.13	2	21.38	2	16.71	2
Italy	0.63	12	0.98	6	0.91	10	0.95	10	1.73	6	1.35	5
Luxembourg	0.29	14	0.25	15	0.40	14	0.20	14	0.03	15	–	15
Netherlands	0.81	9	0.74	9	0.90	11	1.21	8	1.11	7	0.89	10
Norway	0.75	10	0.58	11	1.15	5	4.19	1	2.00	4	1.11	8/9
Portugal	0.92	6	4.13	2	0.94	9	0.85	11	6.13	3	8.63	3
Turkey	1.21	4	10.20	1	1.62	2	1.90	3	119.90	1	39.10	1
UK	1.33	3	1.52	4	0.98	8	0.78	12	0.79	10	1.28	6
US	1.39	2	1.31	5	1.26	3	1.07	9	0.95	8	1.11	8/9
Japan	0.25	15	0.32	14	0.21	15	0.15	15	0.53	12	0.47	14
Non-US NATO	0.92		0.91		1.09		1.24		1.16		1.04	
Non-US NATO + Japan	0.76		0.79		0.87		0.96		0.70		0.93	
Total NATO	1.13		1.09		1.16		1.17		1.06		1.07	
Total NATO + Japan	1.00		1.00		1.00		1.00		1.00		1.00	

*See note, Table A. ADE = Armoured Division Equivalent

Notes

Chapter 1

1 The one exception in the nuclear field being the NATO INF (intermediate nuclear force) decision of December 1979, in which there was active participation by a number of allies.

Chapter 2

1 Secretary of State Dean Acheson testifying before the Senate Foreign Relations Committee, 1949.
2 It was later revealed in Congressional hearings that the substantial increases would amount to four divisions in addition to the two 'occupation' divisions already there.
3 See US Library of Congress, Congressional Research Service, *The Mansfield Proposals to Reduce US Troops in Western Europe 1967-1973* (Washington D.C., US Government Printing Office, 25 June 1980).
4 The Senate also passed an amendment offered by Senator McClellan that 'no ground troops in addition to such four divisions should be sent to Western Europe in implementation of Article 3 of the North Atlantic Treaty without further Congressional approval.'
5. Statement before the Senate Foreign Relations and Armed Services Committees on 8 August 1949 in *Department of State Bulletin*, vol. 21, 22 August 1949, p. 266.
6 General Omar Bradley in a statement before the House Committee on Foreign Affairs. *Hearings on the Mutual Defence Assistance Act of 1949*, 81st Congress, 1st Session.
7 See Robert E. Osgood, *The Entangling Alliance* (University of Chicago Press, 1962), p. 70.
8 Lord Ismay, *NATO—The First Five Years, 1949-1954* (North Atlantic Treaty Organization, 1955), p. 32.
9 NATO Information Service, *NATO: Facts and Figures* (Brussels, 1981), p. 36.

Notes

10 A.C. Enthoven and K. Smith, *How Much is Enough?* (New York, Harper and Row, 1971).
11 The preamble read:

> 'The condition of our European allies, both economically and militarily, has appreciably improved since large contingents of forces were deployed;
> 'The commitment by all Members of the North Atlantic Treaty is based upon the full cooperation of all Treaty partners in contributing materials and men on a fair and equitable basis, but such contributions have not been forthcoming from all other Members;
> 'Relations between the two parts of Europe are now characterized by an increasing two-way flow of trade, people and their peaceful exchange; and
> 'The present policy of maintaining large contingents of US forces and their dependants on the US continent also contributes further to the fiscal and monetary problems of the US.'

12 In 1973 the Senate actually approved the Mansfield Amendment by a close vote of 49 to 46. Taken by surprise, the Administration lobbied for support to defeat the next scheduled amendment, by Senator Cranston, which was identical to that of Senator Mansfield. The Cranston proposal was defeated by a 44 to 51 vote.
13 Robert W. Komer, 'The Trick is How to Get it', *Armed Forces Journal*, August 1981.
14 It is interesting to note that this period of the 1970s, which American analysts have a habit of referring to as the decade when the West slackened off its defence effort because of détente, was a period in which several European nations achieved a substantial degree of modernization, particularly the Bundeswehr.
15 The Long-Term Defence Programme has now been merged into NATO's force-planning cycle.

Chapter 3

1 In order to gain critical votes in favour of the ratification of SALT II, the Carter Administration had to agree to increase defence spending so as to rebuild America's military strength.
2 For an excellent summary of the problems facing the RDF, see Michael Gordon, 'The Rapid Deployment Force—Too Large, Too Small or Just Right for the Task?', *National Journal*, 13 Mar. 1982.
3 France has 4,500 men stationed at Djibouti, 3,200 at Réunion, 350 at Mayotte, and small detachments on three Mozambique islands. The French deploy maritime patrol aircraft to the region, but lack tactical air capabilities.
4 In the case of Greece and Spain, this permission will be even less likely under Socialist governments.

Notes

Chapter 4

1 David Greenwood, 'NATO's 3 per cent solution', *Survival*, Nov./Dec. 1981.
2 Views on the value of the 3 per cent remain mixed. Some officials believe that it was of negative value because it encouraged 'finger-pointing'; others, however, maintain that it was a useful lever for keeping countries up to the mark, and that without it many countries would have allowed their expenditure to decline even further. The latter argument clearly has some merit, but it does not answer Greenwood's argument about whether this really is the most sensible way of obtaining improvements in defence effort.
3 It is worth noting that the force goals established by NATO's military authorities for the period 1981-6 called for countries to accept increases in defence spending of approximately 4 per cent. These force goals were evolved prior to the Soviet invasion of Afghanistan. See *NATO after Afghanistan*, report prepared by the author for the House Committee on Foreign Affairs, 1980.

Chapter 5

1 If fully carried out, the Reagan defence plans for 1981-5 would boost defence spending from about 5½ per cent of GNP to a little under 7½ per cent; see Charles L. Schultze, 'We're Moving Too Fast on Defense', *Los Angeles Times*, 20 Oct. 1982. By 1984, the defence share of the national budget will have risen to 31 per cent (from 24 per cent in 1981); see J. Flint, 'Guns or Butter' *Forbes*, 13 Sept. 1982.
2 Secretary of Defense Weinberger, US Department of Defense, *Annual Defense Department Report 1983*.
3 Public opinion polls indicate that public support for increased defence spending has fallen from its peak of 56 per cent in December 1979 to 29 per cent in 1982; but only 34 per cent favour reductions. Thus a consensus remains for the current level of defence expenditure, which equals 6.3 per cent of GNP; see 'Defending Defence', *National Journal*, 8 Jan. 1983.
4 For details, see US Department of Defense, *Annual Defense Department Report 1983*.
5 Ibid.
6 Senate Armed Services Committee, testimony to the Subcommittee on Sea Power and Force Projection, 9 Mar. 1981.
7 See *The New York Times*, 25 Oct. 1982.
8 See *Washington Post*, 2 June 1982.
9 See, for example, the comments of Senator Nunn that there was 'a huge gap between the rhetoric and reality' of Administration strategy, *Baltimore Sun*, 9 Mar. 1982; also R. W. Komer, 'Maritime Strategy and Coalition Defence', *Foreign Affairs*, Summer 1982.

Notes

10 Much of this section is based on interviews conducted by the author in Washington, D.C., Jan.–Feb. 1982.
11 Earl C. Ravenal, 'Anatomy of the Defence Budget', *Chicago Tribune*, 10 May 1982. In presenting his argument, Ravenal calculated the regional attributions of the 1983 budget as Europe $129 billion, Asia $39 billion, and strategic reserve $36 billion, and commented that, 'given a reasonable projection of current cost growth during the decade to 1992, Europe will cost us $24 trillion'.
12 Jeffrey Record, 'Pull US Troops out of Europe', *US News and World Report*, 1982. See also his 'Should America pay for Europe's Security?', *Washington Quarterly*, Spring 1981, where he shows yet again how burden-sharing criteria can be used selectively to prove that one side is not doing enough. One of the statistics he selects to prove his case is that the US maintains more men under arms as a percentage of national population than most of its key allies in Europe. In doing so, he refers only to active forces, thereby ignoring the dependence on reserve forces inherent in Europe's conscript forces. In fact, when the ratio of active forces to the population is measured, the US ranks third in NATO (behind Greece and Portugal), and when reserve forces are added, it ranks ninth (Norway heading the list).
13 Secretary of Defense Weinberger, interview in *Defence*, December 1981.
14 Secretary of Defense Weinberger, Annual Posture Statement for FY 1983.
15 Richard Perle, Deputy Assistant Secretary of Defense, briefing to the North Atlantic Assembly's Subcommittee on Defence Cooperation, January 1982.
16 As note 13.
17 Remarks to the Wehrkunde Conference, Munich, February 1981.
18 Secretary of Defense Weinberger at the ministerial meeting of the Nuclear Planning Group, Bonn, April 1981.
19 For example, the voting of an amendment to the FY 1982 Authorization Act which contained restrictive measures seriously hindering alliance cooperation in defence procurement; see p. 90.
20 *NATO – Can the Alliance by Saved?* Report of Senator Sam Nunn to the Senate Committee on Armed Services, 13 May 1982.
21 Interviews conducted by the author, Washington, D.C., Jan.–Feb. 1982.
22 The proposal to deploy the ERW in Europe was particularly significant because it drew public attention to the existence of battlefield nuclear weapons, systems with which the public could easily identify because of their location and because they were a very real and 'usable' component of NATO strategy.
23 For a detailed discussion of the gas pipeline dispute and of US–European trade relations, see Stephen Woolcock, 'US–European Trade Relations', *International Affairs*, Autumn 1982.

Notes

24 The debatable legal basis for the sanctions is examined in Douglas E. Rosenthal and William M. Knighton, *National Laws and International Commerce: The Problem of Extraterritoriality*, Chatham House Paper No. 17 (London, Routledge & Kegan Paul, 1982).
25 Declaration of the Heads of State and Government participating in the meeting of the North Atlantic Council in Bonn, 10 June 1982.
26 German economists have estimated that depreciation of the German mark against the US dollar by one pfennig, forces an unavoidable increase in German defence spending (primarily on fuel, but also for US equipment) of 4 million DM.
27 Senator Ted Stevens' statement to the US Senate, *Congressional Record*, 27 Sept. 1982.
28 For details, see *Congressional Record*, 16 Mar. 1982, pp. H895–902.
29 Despite widespread criticism of the Allies' performance, the House of Representatives firmly rejected the idea of troop withdrawals by voting down by 314 to 87 an amendment by Representative Pat Schroeder that would have cut the number of Americans stationed abroad by half (see *Congressional Record*, 29 July 1982).
30 Congressional Research Service, Foreign Affairs and National Defence Division, *Crisis in the Atlantic Alliance: Organs and Implications*, April 1982.
31 *NATO Today: the Alliance in Evolution*, a report to the Senate Committee on Foreign Relations, April 1982.
32 Nunn, op. cit.
33 For detailed discussion of this issue, see *The Implications of Technology for the Battlefield*, a report presented by Mr Patrick Duffy on behalf of the Subcommittee on Defence Cooperation of the Military Committee of the North Atlantic Assembly, October 1982.
34 Senator Ted Stevens, *Congressional Record*, 13 May 1982, p. S5193.

Chapter 6

1 *NATO Financial and Economic Board Report*, 1951.
2 *Report on Allied Contributions to the Common Defense*, a report to the US Congress by Caspar W. Weinberger, Secretary of Defense (Washington, D.C., Department of Defense, March 1982).
3 A number of countries which operate conscription on a selective basis — that is, they conscript a certain proportion of available youth — actually pay their conscripts relatively high wages in order to compensate for the unfairness of being selected. Thus, in comparisons of this nature, several countries such as Canada, Luxembourg and the Netherlands, whose pay and allowances are higher than in the US, would have their defence expenditure adjusted downward. See *Report on Allied Contributions to the Common Defense*, op. cit.
4 Ibid.

Notes

Chapter 7

1. Declaration of the Heads of State and Government, North Atlantic Council, Bonn, 10 June 1982.
2. 'Document on Integrated NATO Defence', Bonn, 10 June 1982; reproduced in *NATO Review*, vol. 30 (1982), no. 3.
3. McGeorge Bundy, George Kennan, Robert McNamara and Gerard Smith, 'Nuclear Weapons and the Atlantic Alliance', *Foreign Affairs*, Spring 1982.
4. Karl Kaiser, George Leber, Alois Mertes and Franz-Josef Schulze, 'Nuclear Weapons and the Preservation of Peace', *Foreign Affairs*, Summer 1982.
5. Nunn, op. cit. (ch. 5, n. 20).
6. 'Bundestag Member Urges Conventional Force Change', *Aviation Week and Space Technology*, 19 July 1982.
7. 'Do You Sincerely Want To Be Non-nuclear?', *The Economist*, 31 July 1982.
8. Edward N. Luttwak, 'How to Think about Nuclear War', *Commentary*, August 1982.
9. Many of the ideas in Luttwak's critique are consistent with a school of thought which advocates that NATO should move from its current forward defence, or attrition strategy, to defence in depth or a manoeuvre strategy. For a thorough discussion of these ideas, see John J. Mearsheimer, 'Manpower, Mobile Defence and the NATO Central Front', *International Security*, Winter 1981/2; also the writings of Steven Canby, esp. 'Military Reform and the Art of War', *International Security Review*, vol. 7 (1982), no. 3.
10. *The Airland Battle and Corps 86*, US Army Training and Doctrine Command pamphlet, 25 March 1981.
11. Nunn, op. cit.
12. For a discussion of the development of European Political Cooperation, see Christopher Hill, ed., *National Foreign Policies and European Political Cooperation*, to be published by Allen and Unwin in autumn 1983.
13. Christopher Tugendhat, 'Europe's need for Self-Confidence', *International Affairs*, vol. 58, no. 1.
14. See, for example, 'The European Community—Past, Present and Future', a speech by President Dankert at St Catherine's College, Oxford.

For Product Safety Concerns and Information please contact our EU representative GPSR@taylorandfrancis.com
Taylor & Francis Verlag GmbH, Kaufingerstraße 24, 80331 München, Germany

www.ingramcontent.com/pod-product-compliance
Lightning Source LLC
Chambersburg PA
CBHW052135300426
44116CB00010B/1905